AF725327

Addition Properties

Property	In Symbols	Examples
Closure	$a + b$ is unique	$2 - 4 = -2$ $\frac{1}{2} + \frac{7}{8} = 1\frac{3}{8}$
Commutative	$a + b = b + a$	$14 + 16 = 16 + 14$
Associative	$(a + b) + c = a + (b + c)$	$\left(\frac{1}{8} + \frac{1}{4}\right) + \frac{1}{2} = \frac{1}{8} + \left(\frac{1}{4} + \frac{1}{2}\right)$
Identity	$a + 0 = 0 + a = a$	$-7.3 + 0 = 0 + (-7.3) = -7.3$
Inverse	$a + (-a) = 0$	$13.6 + (-13.6) = 0$ $\frac{1}{2} + \left(-\frac{1}{2}\right) = 0$

Multiplication Properties

Property	In Symbols	Examples
Closure	ab is unique	$(-5)\cdot(4) = -20$ is unique
Commutative	$ab = ba$	$25 \cdot 5 = 5 \cdot 25$
Associative	$(ab)c = a(bc)$	$\left(\frac{3}{8}\cdot\frac{2}{5}\right)\cdot\frac{10}{9} = \frac{3}{8}\cdot\left(\frac{2}{5}\cdot\frac{10}{9}\right)$
Identity	$a \cdot 1 = 1 \cdot a = a$	$26 \cdot 1 = 1 \cdot 26 = 26$
Property of Zero	$a \cdot 0 = 0 \cdot a = 0$	$83.2 \cdot 0 = 0 \cdot 83.2 = 0$
Inverse	$a \cdot \frac{1}{a} = \frac{1}{a} \cdot a = 1,$ $a \neq 0$	$5 \cdot \frac{1}{5} = \frac{1}{5} \cdot 5 = 1$

Problem Solving Steps

READ	Read the entire problem slowly and carefully. Then go back and read the question again.
PLAN	Decide what you can do to solve the problem.
SOLVE	Put your plan into action.
ANSWER	State your answer in terms of the original question.
CHECK	Determine whether or not your answer is *reasonable*.

Other Mathematical Properties

Property	In Symbols	Examples
Distributive	$a(b + c) = ab + ac$ or $a(b - c) = ab - ac$	$3(5 + 4) = 3 \cdot 5 + 3 \cdot 4$ $3(5 - 4) = 3 \cdot 5 - 3 \cdot 4$
Transitive	If $a = b$ and $b = c$, then $a = c$	If $3 + 5 = 4 + 4$ and $4 + 4 = 2 + 6$ then $3 + 5 = 2 + 6$
Reflexive	$a = a$ (any number is equal to itself)	$8 = 8$
Substitution	If $a = b$, then we can substitute b for a or substitute a for b in any statement	If $x = 4$, then $x + 2 = 4 + 2$
Symmetric	If $a = b$, then $b = a$	If $4 + 6 = 5 + 5$, then $5 + 5 = 4 + 6$

Exponents

$$n^1 = n \qquad n^0 = 1$$

$$15^1 = 15 \qquad 15^0 = 1$$

Properties of Exponents for Multiplication

For all real numbers *a* and *b*, if *m* and *n* are integers, then

$$a^m \cdot a^n = a^{m+n} \qquad (a^m)^n = a^{mn} \qquad (ab)^m = a^m b^m$$

Properties of Exponents for Division

For all real numbers *a* and *b*, if *m* and *n* are integers, then

$$\frac{a^m}{a^n} = a^{m-n}, \text{ if } a \neq 0 \qquad \left(\frac{a}{b}\right)^m = \frac{a^m}{b^m}, \text{ if } b \neq 0$$

Scientific Notation

Number	Scientific Notation
2,650,000	2.65×10^{6}
0.0073	7.3×10^{-3}

number greater than or equal to 1 and less than 10

power of 10

Iterations

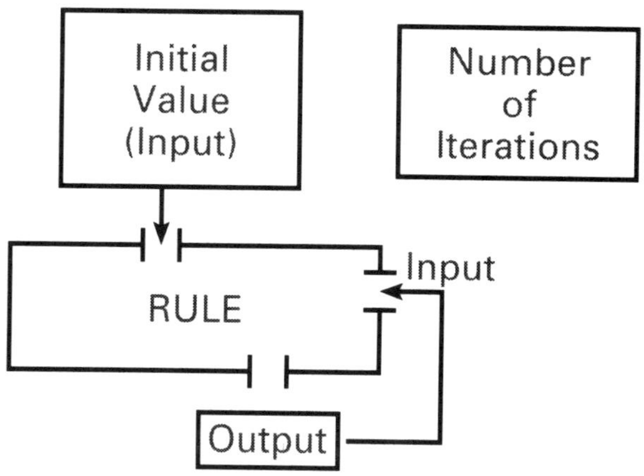
Initial
Value
(Input)
Number
of
Iterations
RULE
Input
Output

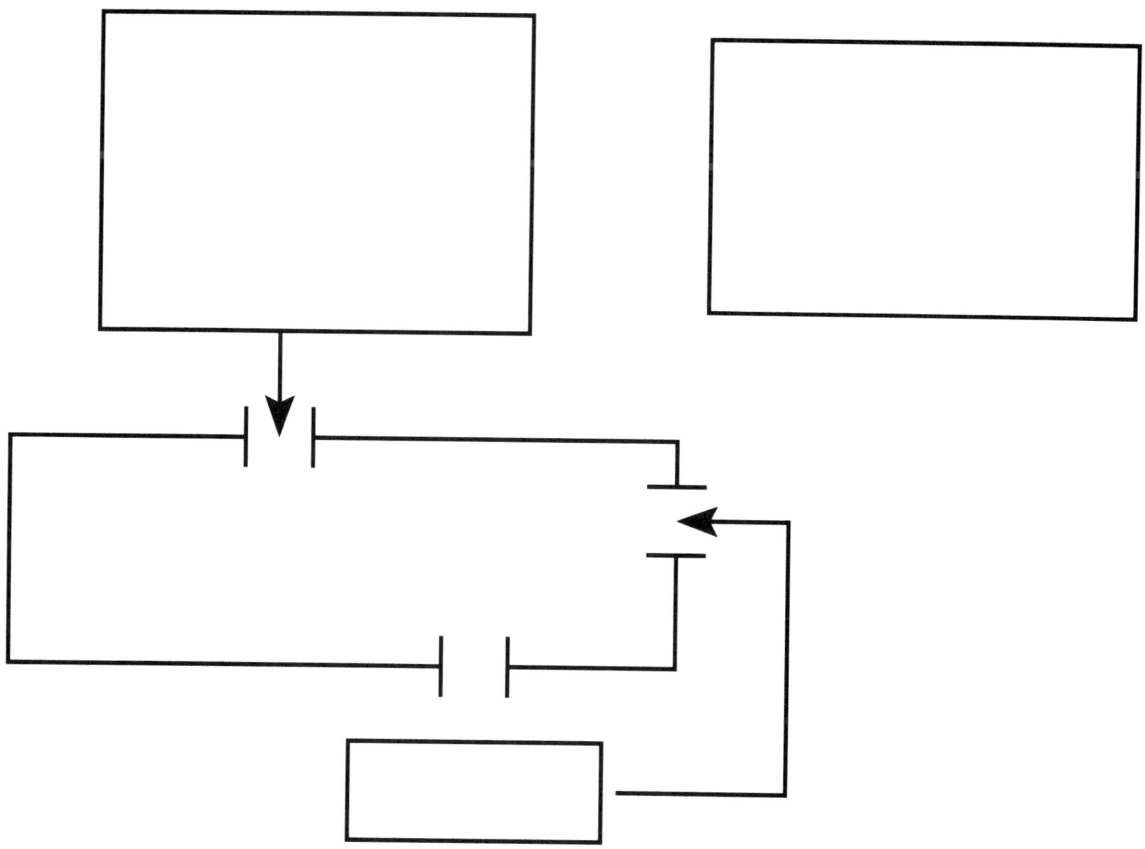

Coordinate Plane A

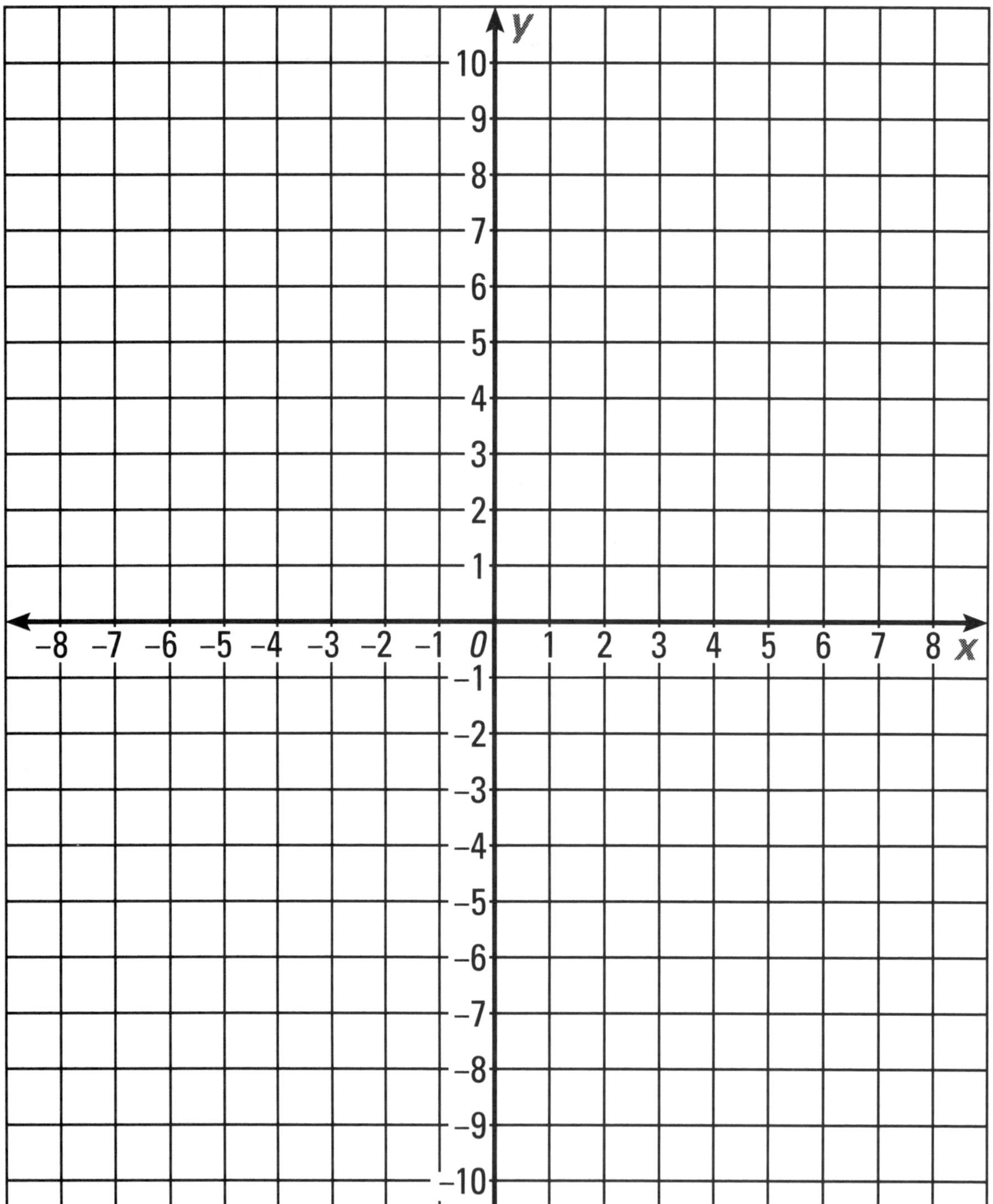

Coordinate Plane B

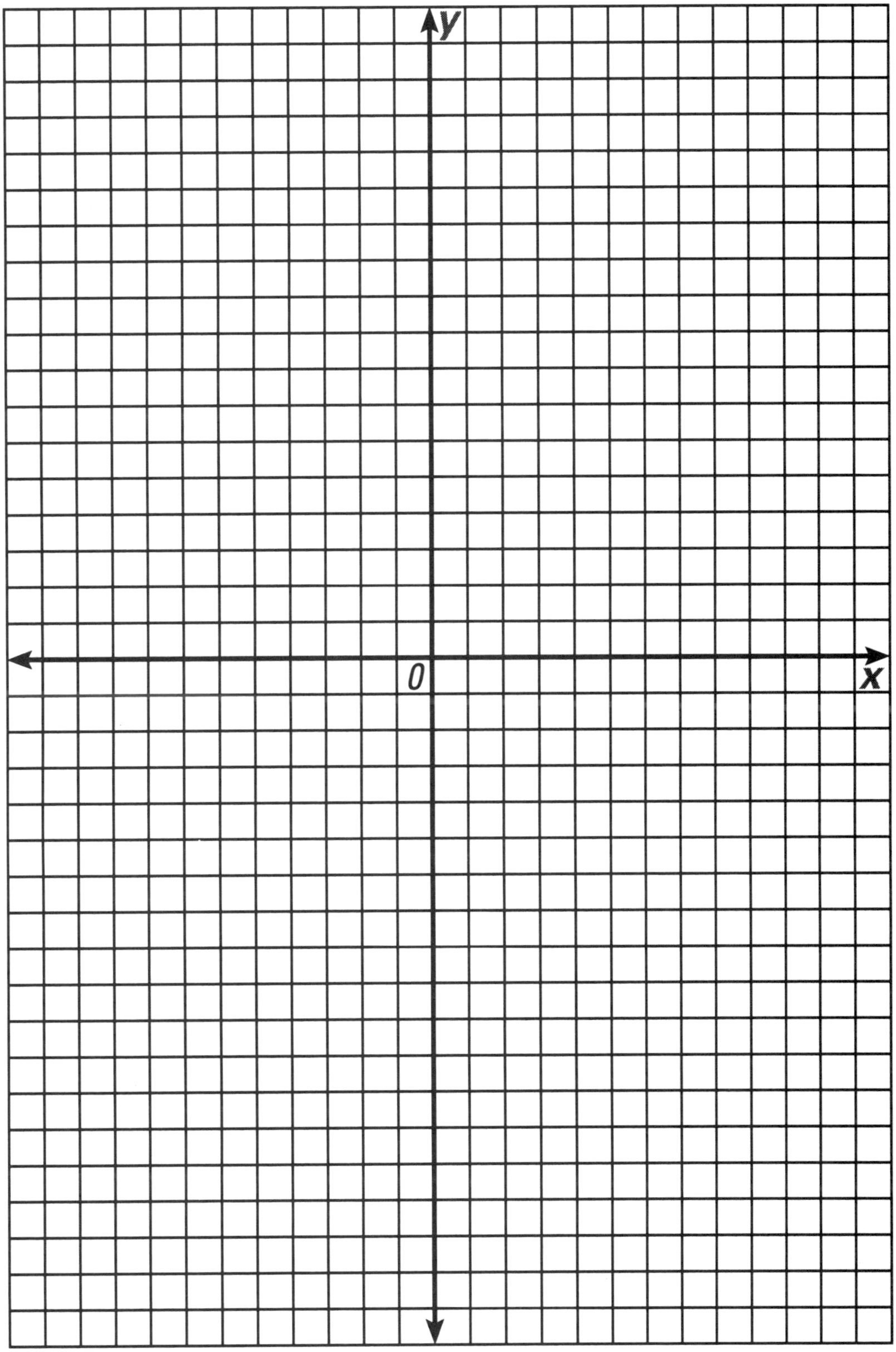

Algeblocks™ Sentences Mat

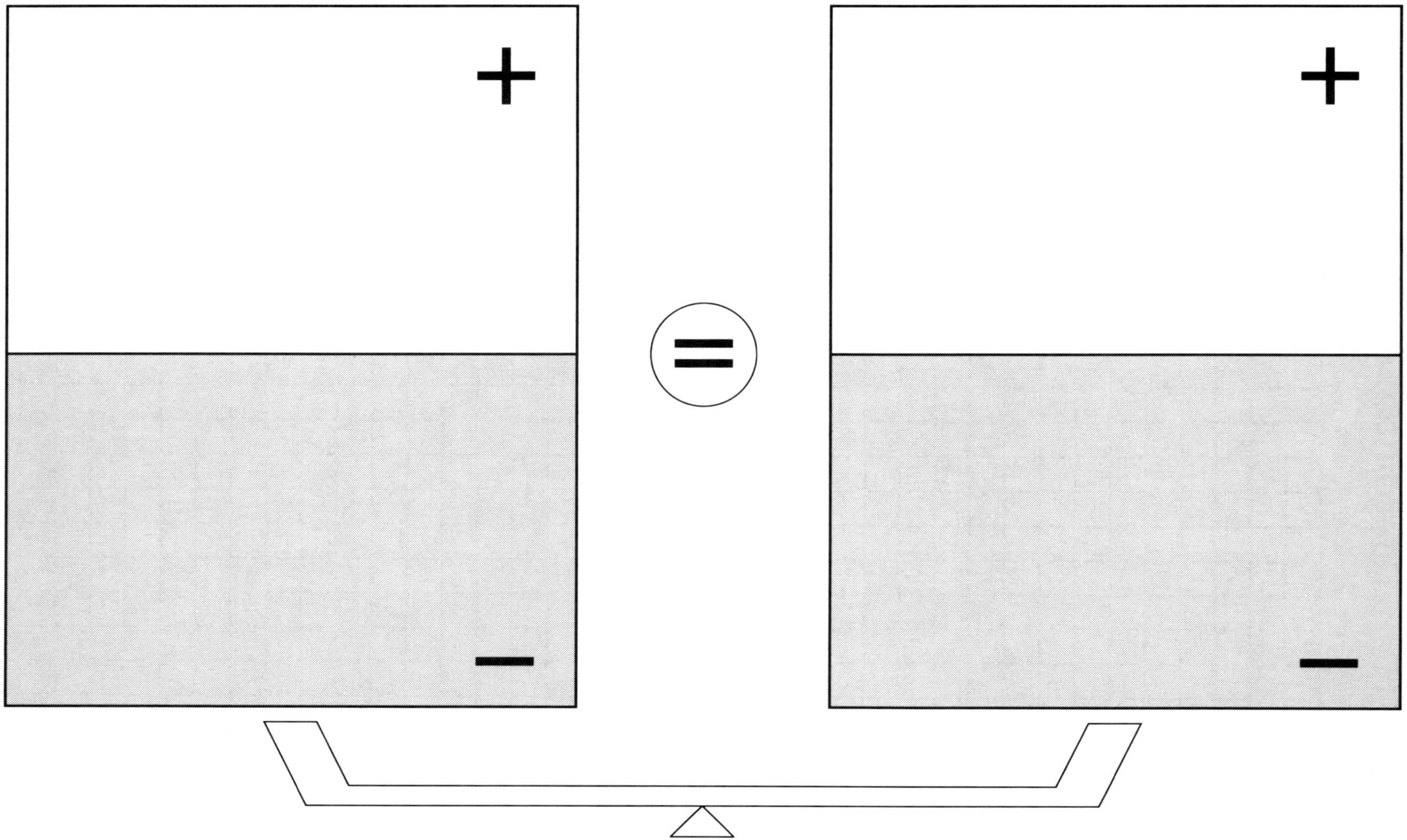

Algeblocks™

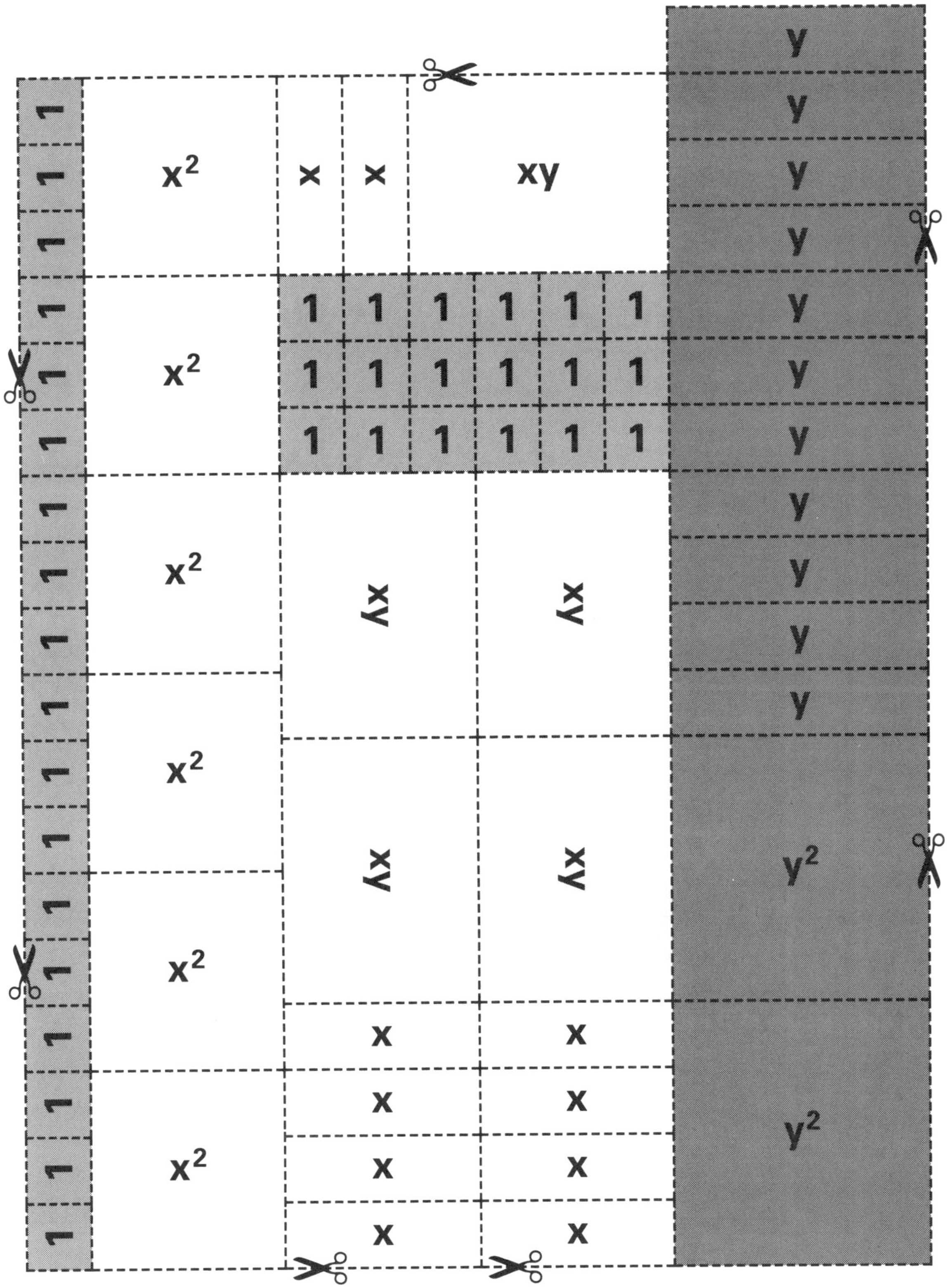

Half-Planes

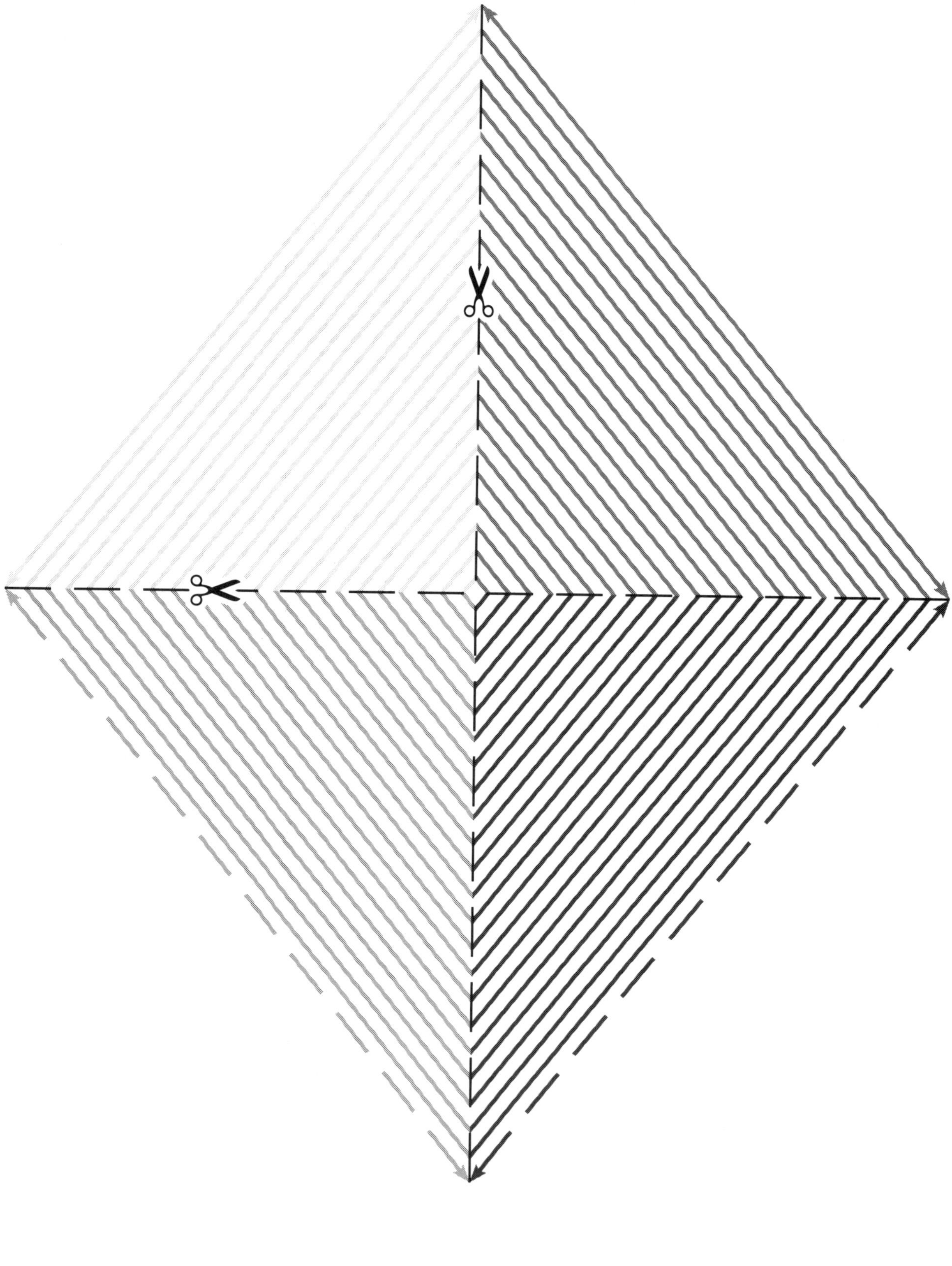

Statistics

- ✔ The **mean**, or arithmetic average, is found by dividing the sum of the data by the number of data.

- ✔ The **median** is the middle number when the data are arranged in numerical order. (If the number of data is even, the median is the average of the *two* middle numbers.)

- ✔ The **mode** is the number that occurs most frequently in the set of data. Some sets of data have no mode; some have more than one mode.

- ✔ The **range** is the difference between the greatest and the least values in a series of data.

Points and Planes

A point is a location in space having no dimensions.

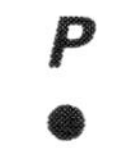

point *P*

A plane contains an infinite set of points and has no boundaries.

plane *M*

Collinear points are points that lie on the same line.

Noncollinear points are points that do not lie on the same line.

Coplanar points are points that lie in the same plane.

Noncoplanar points are points that do not lie in the same plane.

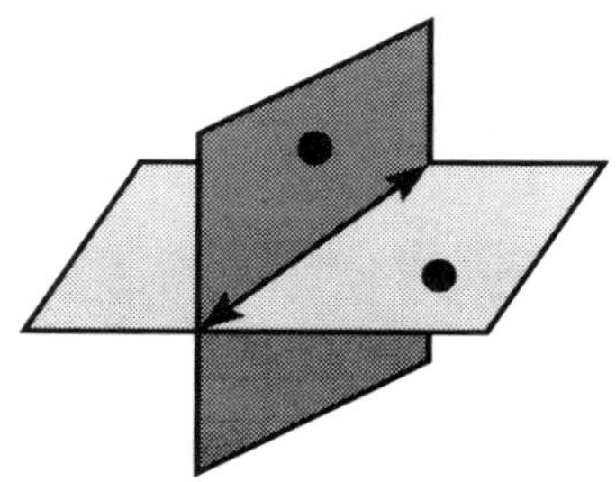

Lines, Segments, and Rays

A **line** is a set of points that extends infinitely in two opposite directions.

line *TB* ($\overleftrightarrow{TB}$), line *BT* ($\overleftrightarrow{BT}$), or line *k*

A **segment (line segment)** is a part of a line consisting of two **endpoints** and all points that lie between them.

segment *NC* ($\overline{NC}$) or segment *CN* ($\overline{CN}$)

A **ray** is part of a line that starts at one endpoint and extends without end in one direction.

ray *SN* ($\overrightarrow{SN}$)

Angles

An angle is the figure formed by two rays that have a common endpoint. The endpoint is called the vertex of the angle.

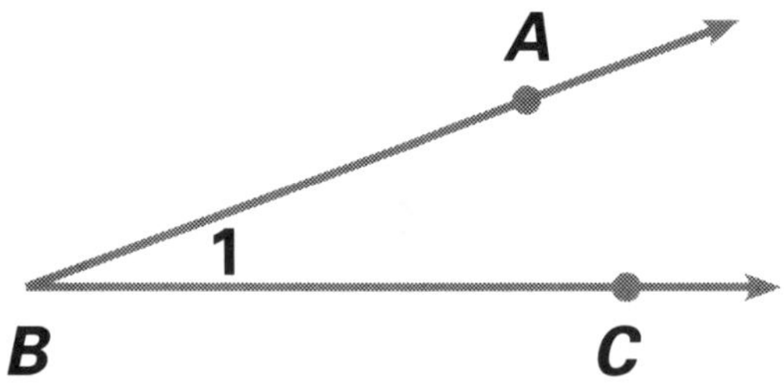

angle *ABC* (∠*ABC*) or
angle *CBA* (∠*CBA*) or
angle *B* (∠*B*) or angle 1 (∠1)

Acute angle	Right angle	Obtuse angle	Straight angle
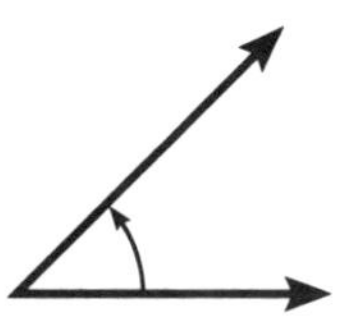	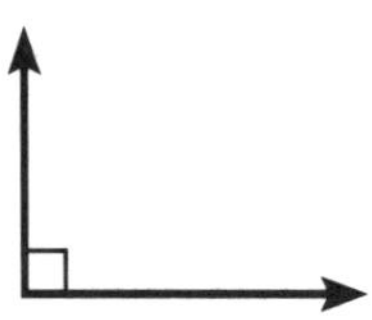	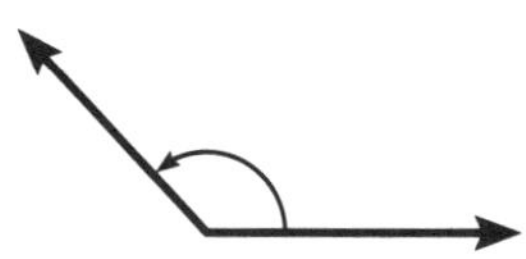	
less than 90°	exactly 90°	greater than 90°, but less than 180°	exactly 180°

More about Angles

Complementary Angles

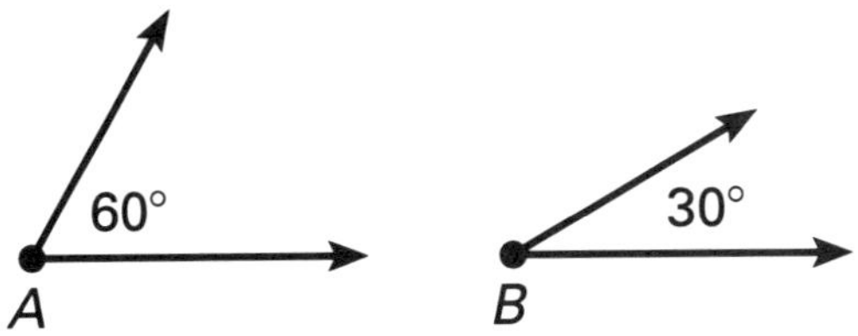

$$m\angle A + m\angle B = 90°$$

Supplementary Angles

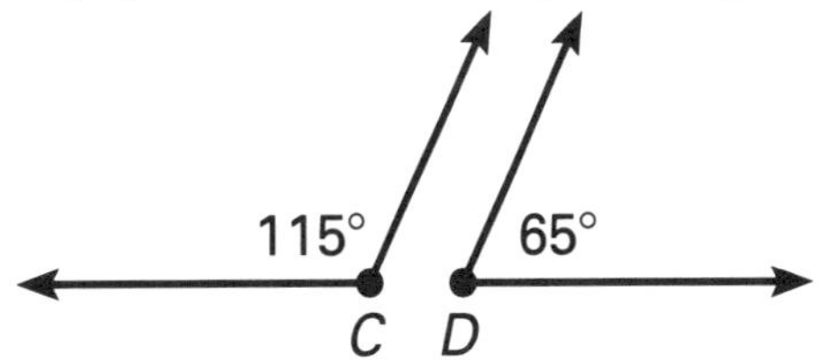

$$m\angle C + m\angle D = 180°$$

Adjacent Angles

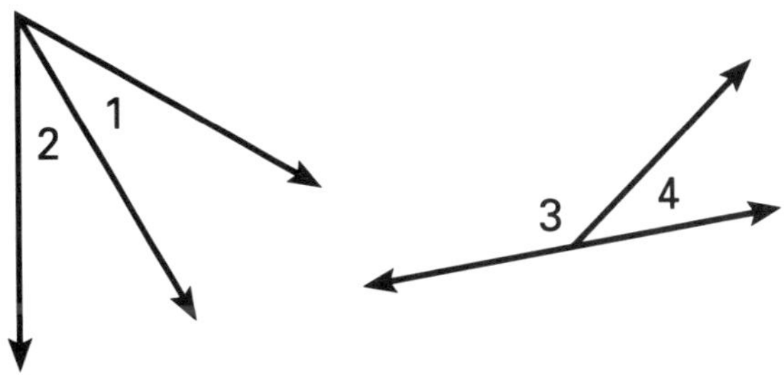

Vertical Angles

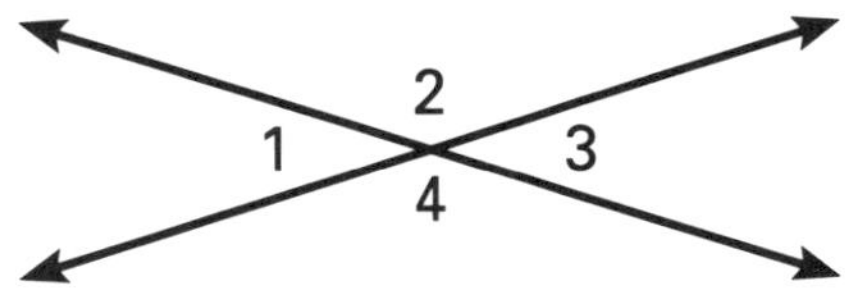

$\angle 1$ and $\angle 3$ are vertical angles.
$\angle 2$ and $\angle 4$ are vertical angles.

Congruent angles have the same measure

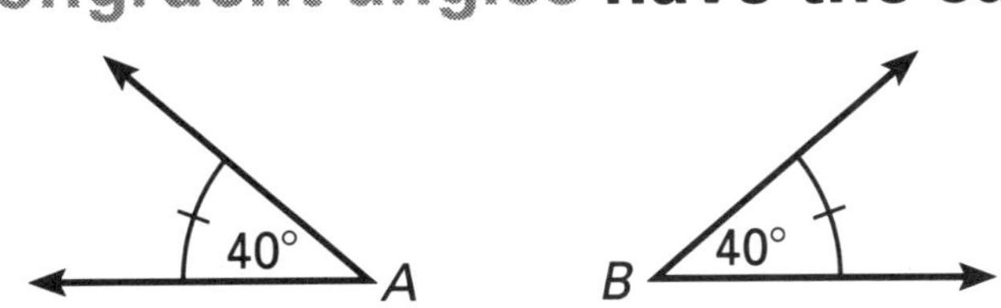

$$\angle A \cong \angle B$$

Interior and Exterior Angles

A transversal is a line that intersects each of two other coplanar lines in different points. The intersection of a transversal with two lines produces interior and exterior angles. In the figure below, transversal *t* intersects lines *m* and *n*.

alternate interior angles:
∠1 and ∠7
∠2 and ∠8

alternate exterior angles:
∠5 and ∠3
∠6 and ∠4

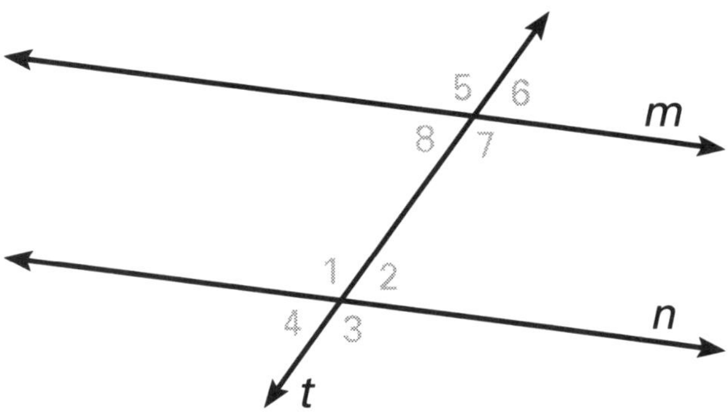

same-side interior angles:
∠2 and ∠7
∠1 and ∠8

corresponding angles:
∠1 and ∠5, ∠2 and ∠6;
∠3 and ∠7, ∠4 and ∠8

Deductive Proof

Given:

Prove:

Statements	Reasons

Classifying Triangles

Classified by angle measure

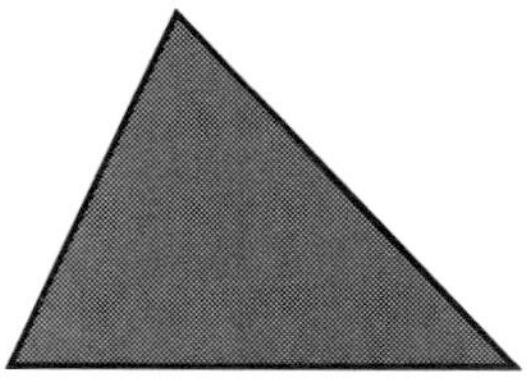

Acute triangle
three acute angles

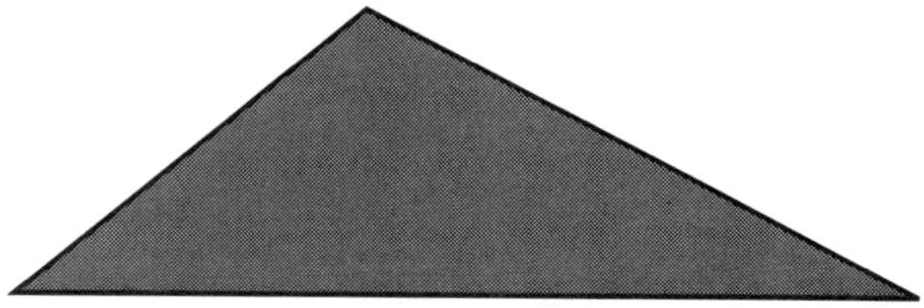

Obtuse triangle
one obtuse angle

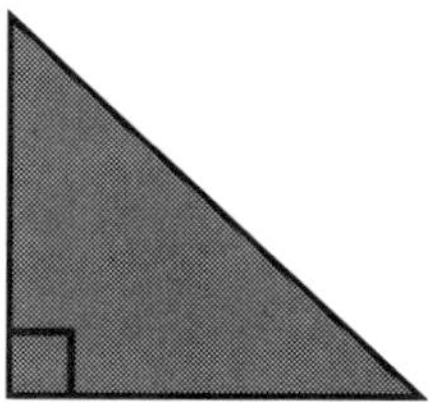

Right triangle
one right angle

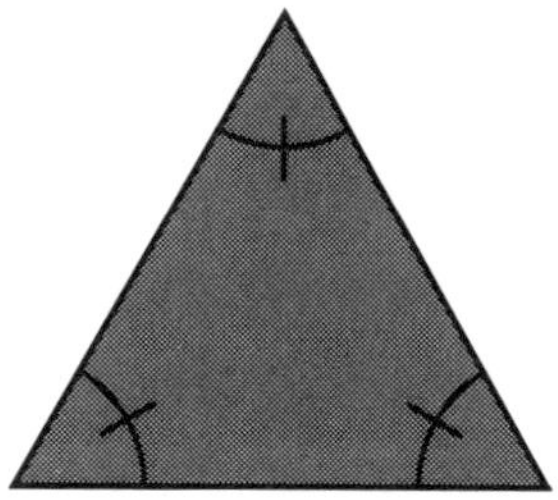

Equiangular triangle
all three angles congruent

Classified by lengths of sides

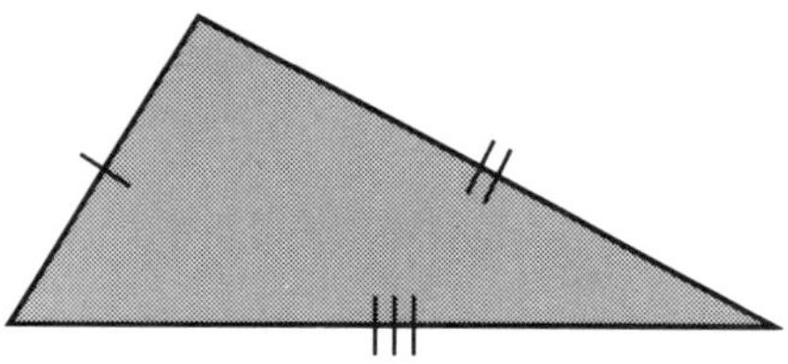

Scalene triangle
no congruent sides

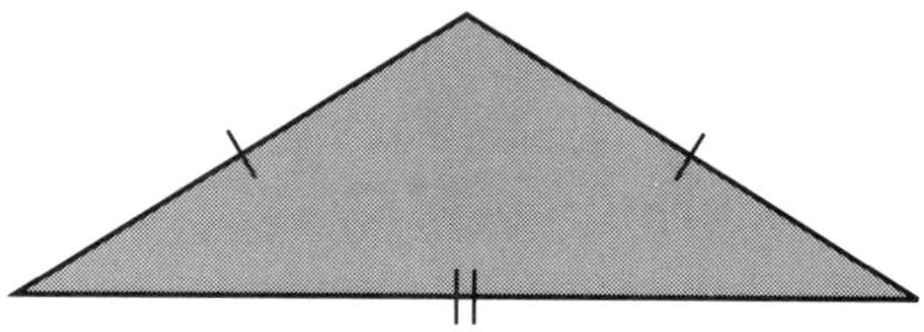

Isosceles triangle
at least two congruent sides

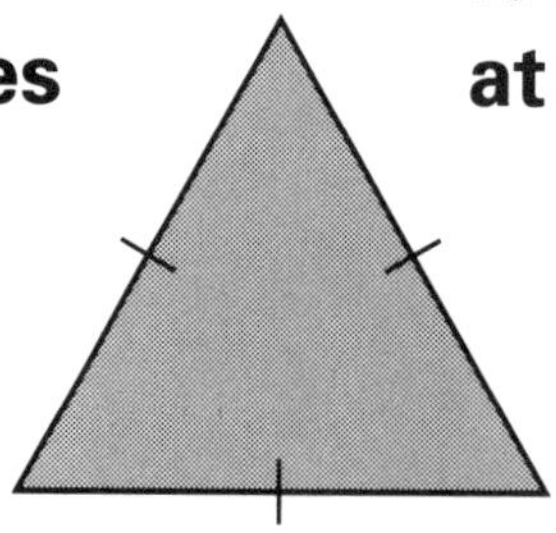

Equilateral triangle
all three sides congruent

Congruent Triangles

Two triangles are congruent if their vertices can be matched so that corresponding parts of the triangles are congruent.

Side-Side-Side Postulate (SSS):

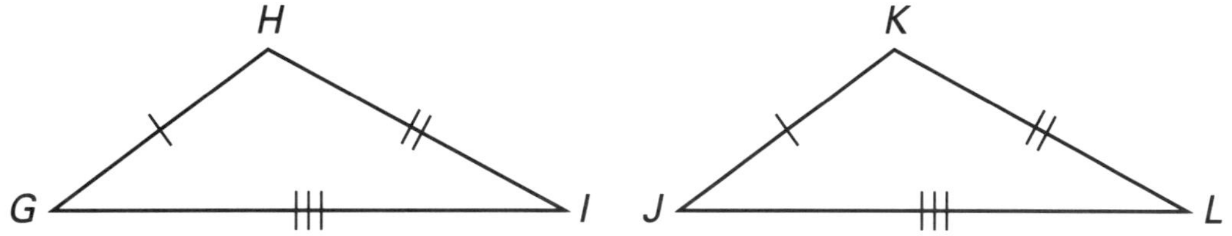

Side-Angle-Side Postulate (SAS):

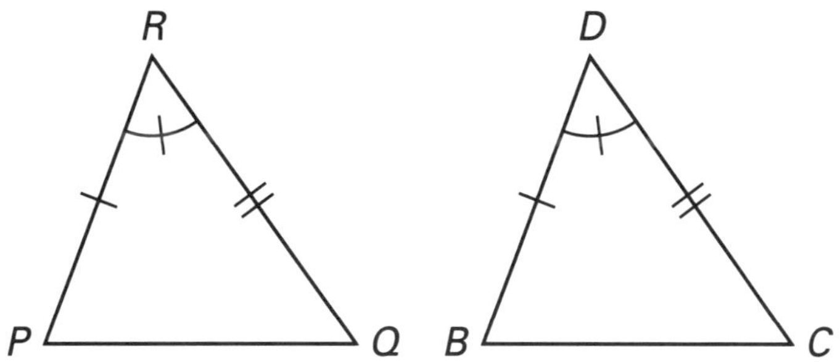

Angle-Side-Angle Postulate (ASA):

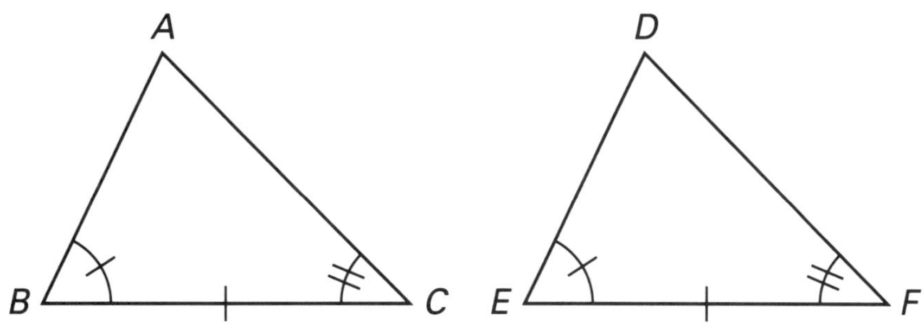

Polygons

convex polygon

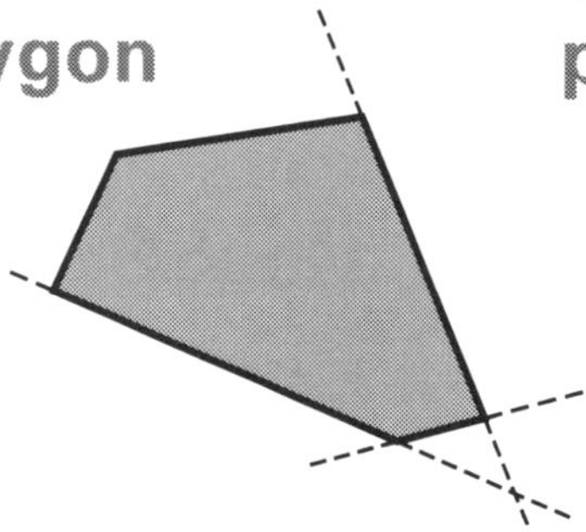

concave polygon

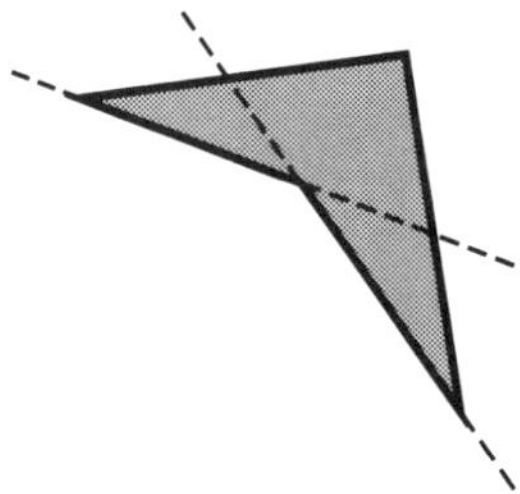

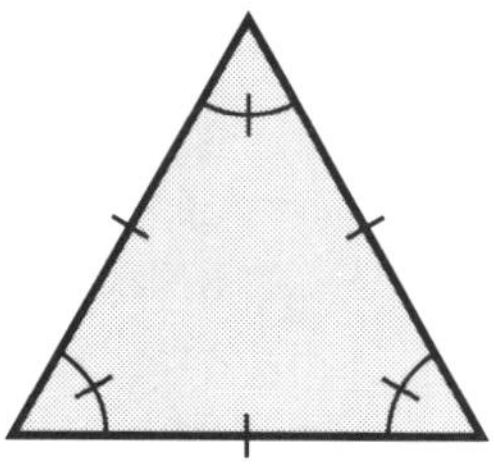

regular triangle

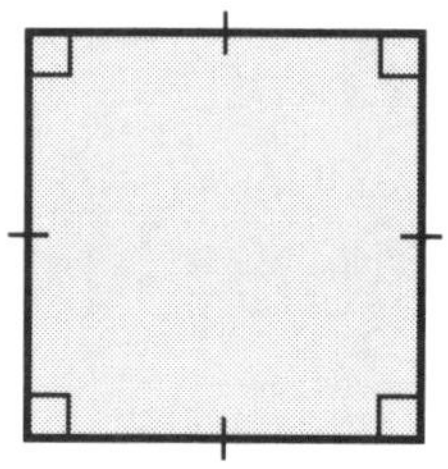

regular quadrilateral

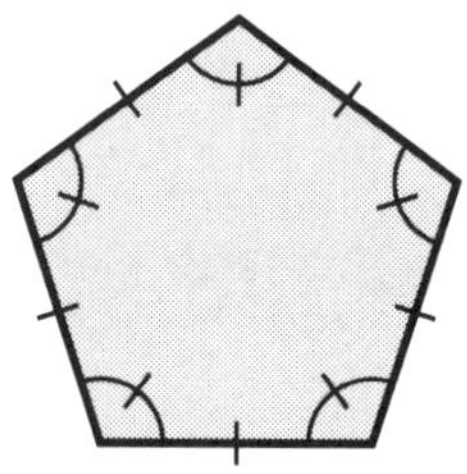

regular pentagon

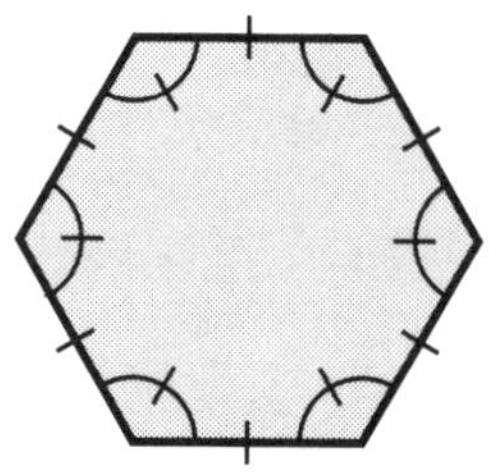

regular hexagon

You can find the sum of the angles of any polygon by drawing the diagonals from any vertex.

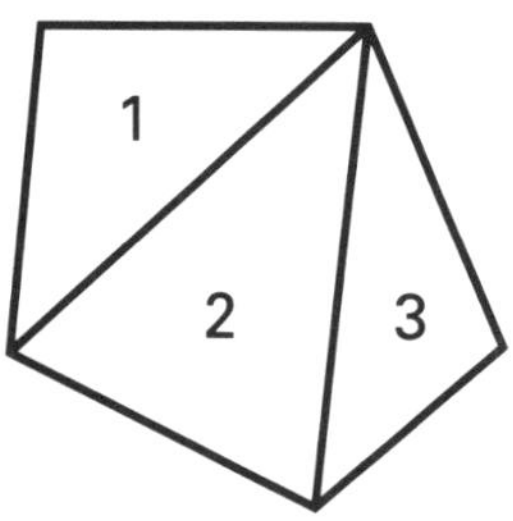

5 sides
3 triangles
sum = 3 × 180°

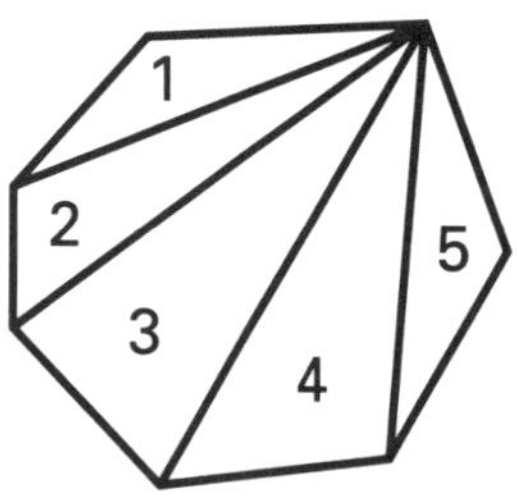

7 sides
5 triangles
sum = 5 × 180°

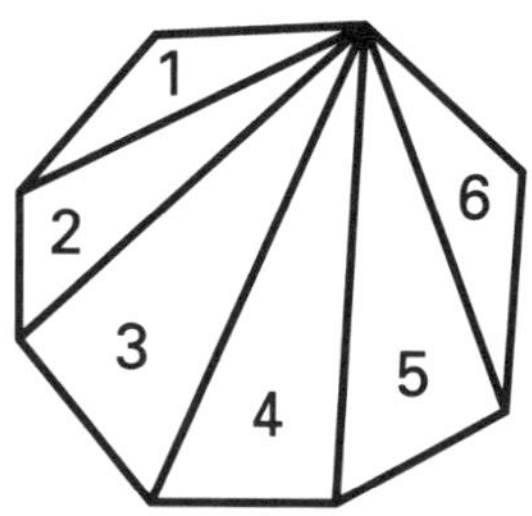

8 sides
6 triangles
sum = 6 × 180°

Quadrilaterals

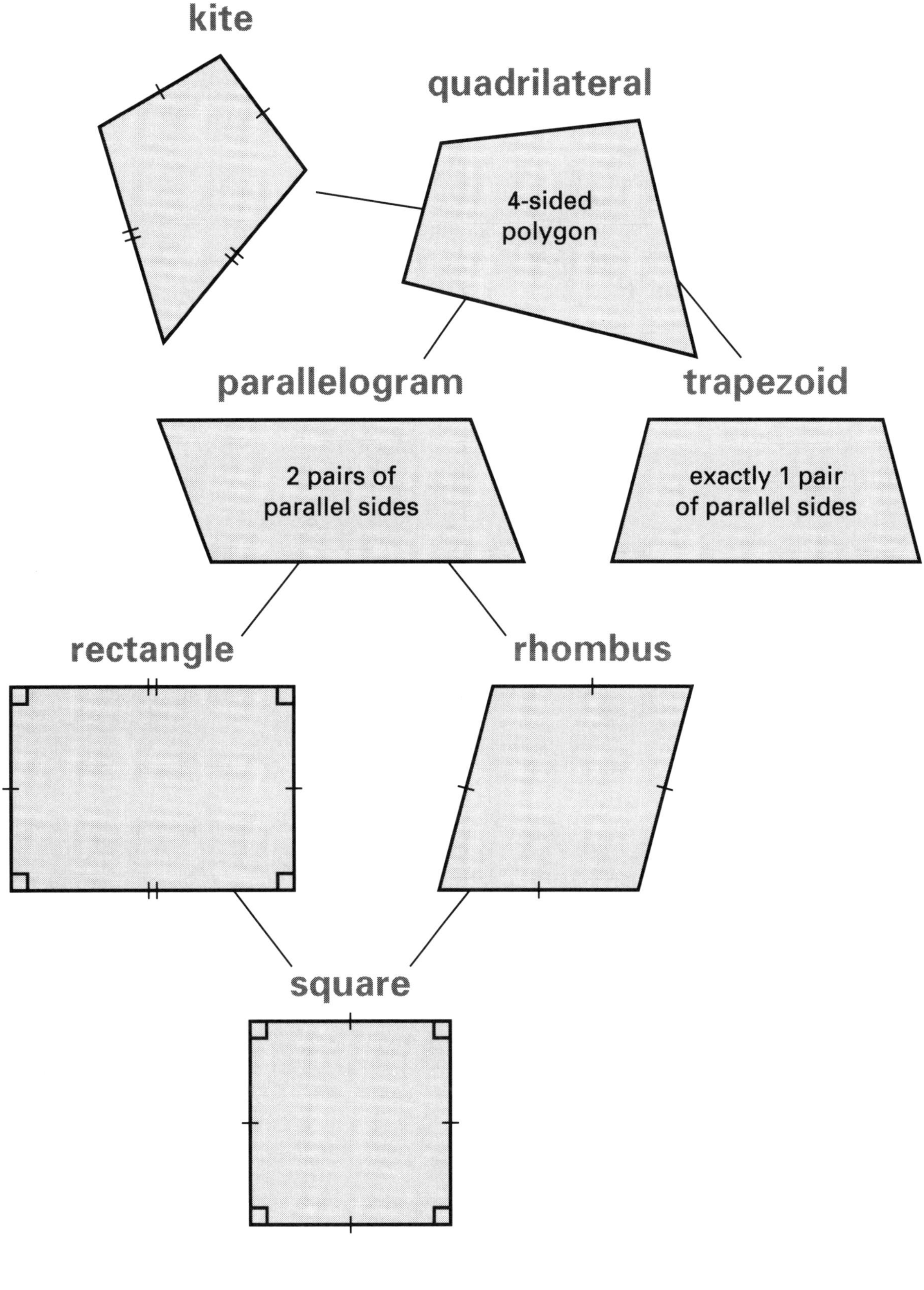

Formulas

l = length	w = width
b = base	h = height
s = side	r = radius
d = diameter	
B = area of the base of a three-dimensional figure	

Perimeter: *P*	Area: *A*
Rectangle: $P = 2l + 2w$	Rectangle: $A = lw$
Square: $P = 4s$	Parallelogram: $A = bh$
Regular Polygon with n sides: $P = ns$	Trapezoid: $A = \frac{1}{2}h(b_1 + b_2)$
Triangle: $P = s_1 + s_2 + s_3$	Triangle: $A = \frac{1}{2}bh$
Circle (Circumference): $C = \pi d$ or $C = 2\pi r$	Circle: $A = \pi r^2$
	Square: $A = s^2$

Surface Area: *SA*	Volume: *V*
Rectangular Prism: $SA = 2(lw + lh + wh)$	Prism: $V = Bh$
Cube: $SA = 6s^2$	Rectangular Prism: $V = lwh$
Cylinder: $SA = 2\pi rh + 2\pi r^2$	Cube: $V = s^3$
Sphere: $SA = 4\pi r^2$	Cylinder: $V = \pi r^2 h$
Cone: $SA = \pi rs + \pi r^2$	Sphere: $V = \frac{4}{3}\pi r^3$
	Pyramid: $V = \frac{1}{3}Bh$
	Cone: $V = \frac{1}{3}\pi r^2 h$

Prisms

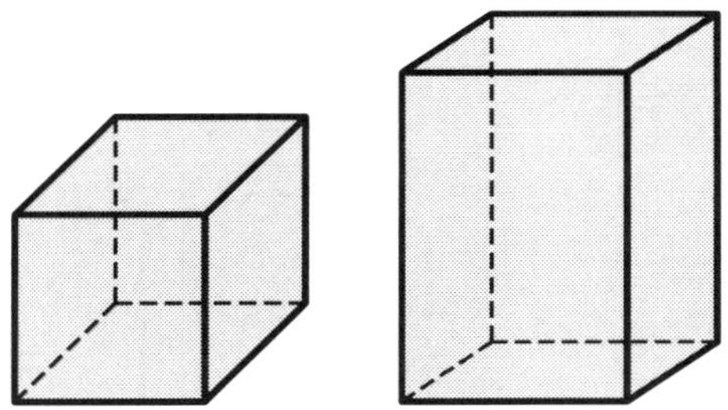

right rectangular prisms

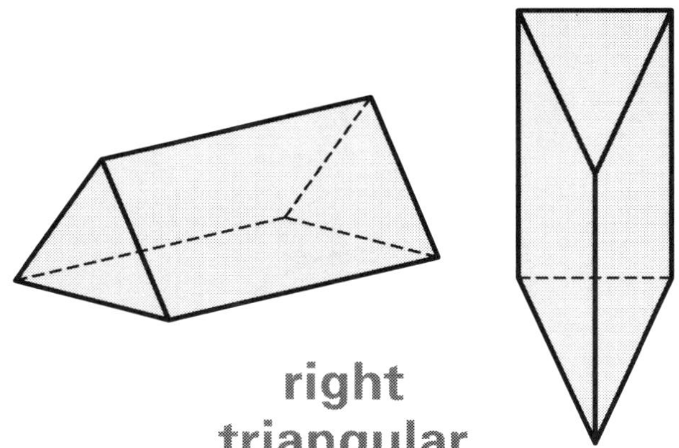

right triangular prisms

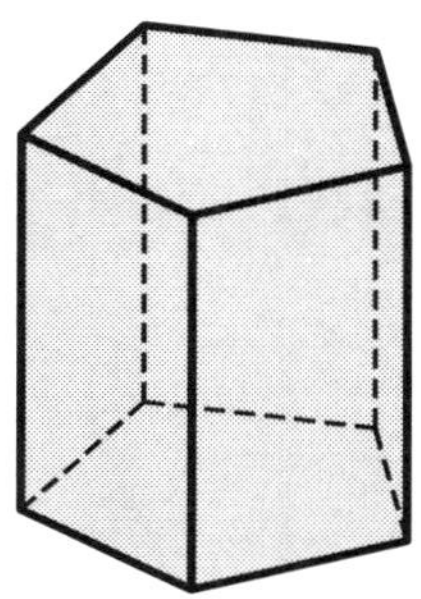

right pentagonal prisms

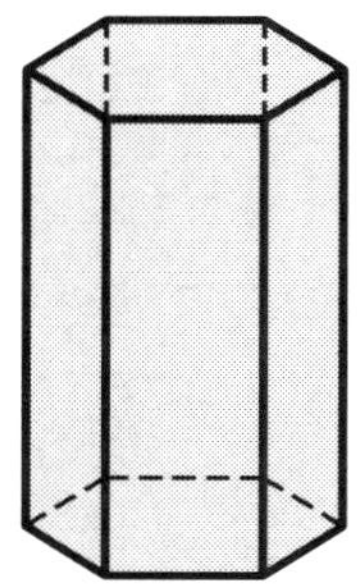

right hexagonal prisms

Pyramids and Other Three-Dimensional Shapes

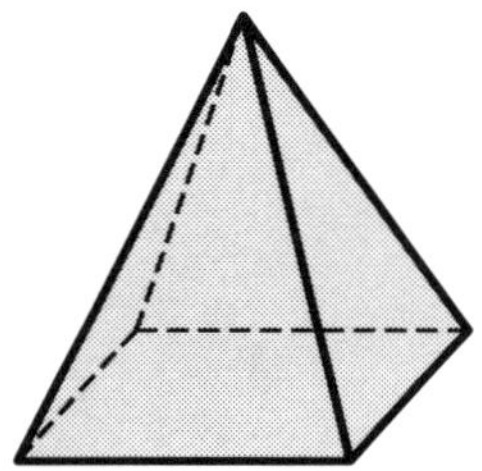

right square pyramid

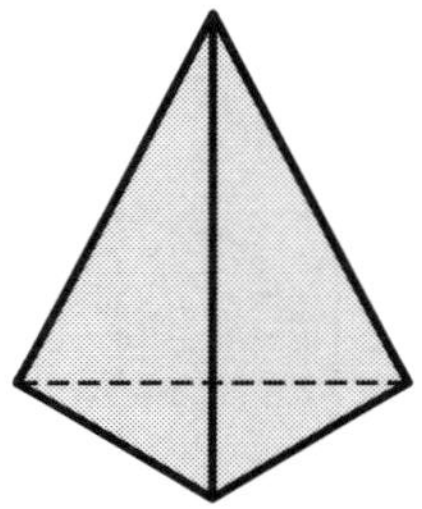

right triangular pyramid

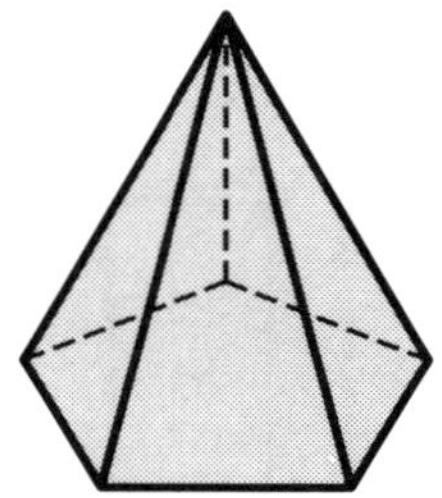

right pentagonal pyramid

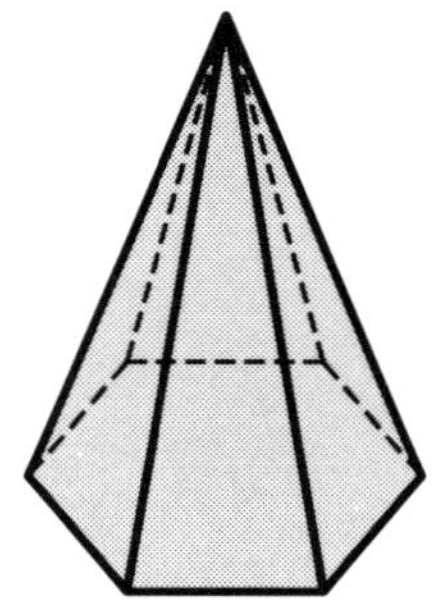

right hexagonal pyramid

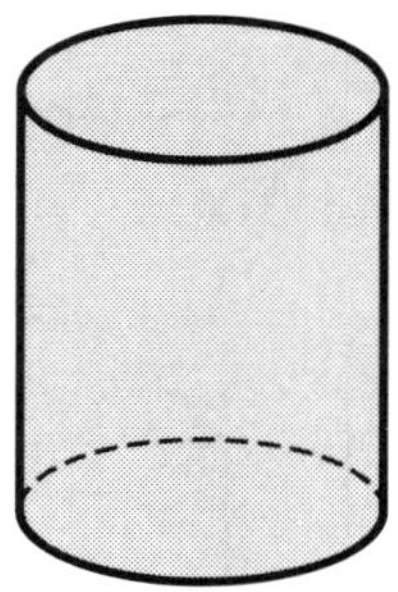

right cylinder

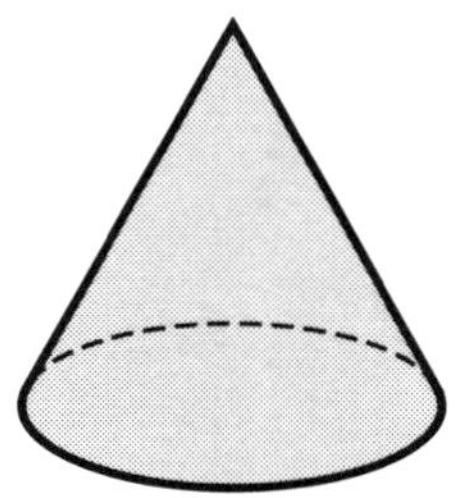

right cone

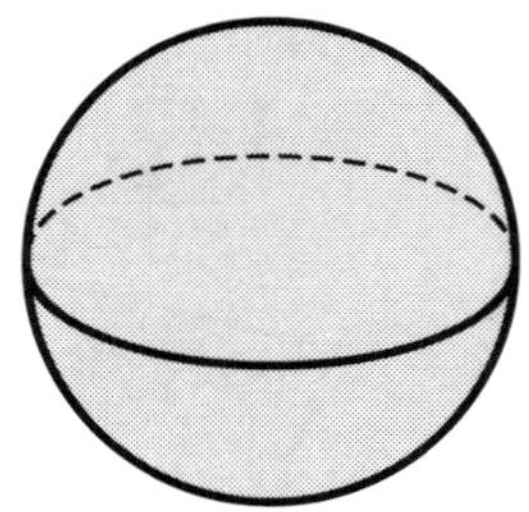

sphere

Slope of a Line

The slope of a segment is the ratio of its *rise* (vertical distance) compared with its *run* (horizontal distance).

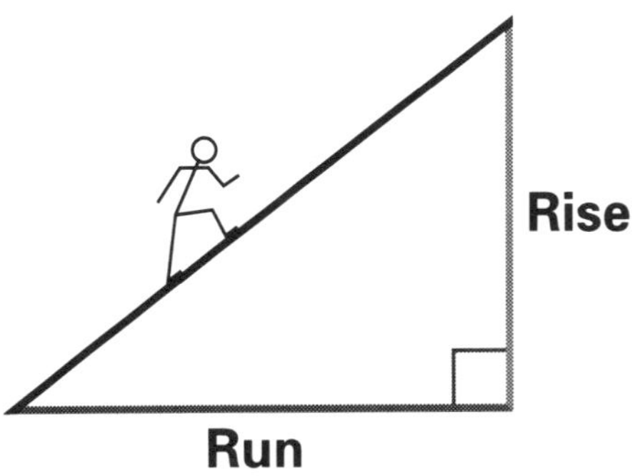

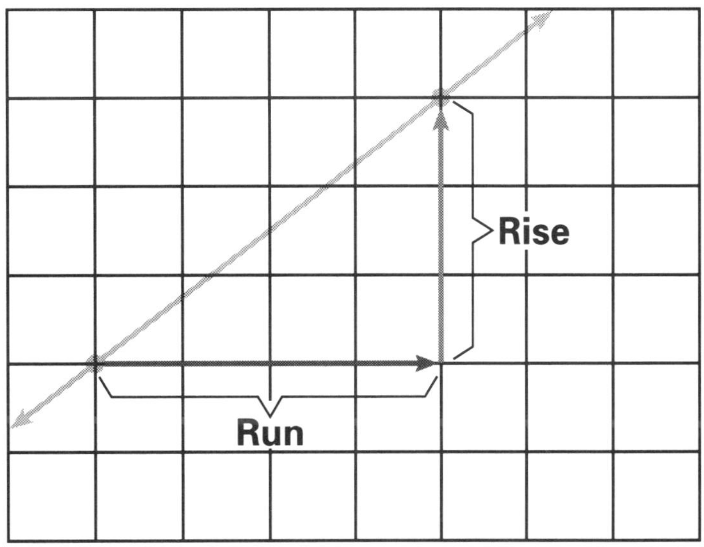

$$\text{slope} = \frac{\text{rise}}{\text{run}} = \frac{\text{change in } y\text{-coordinates}}{\text{change in } x\text{-coordinates}} = \frac{y_2 - y_1}{x_2 - x_1}$$

Slope of $\overleftrightarrow{AB}$

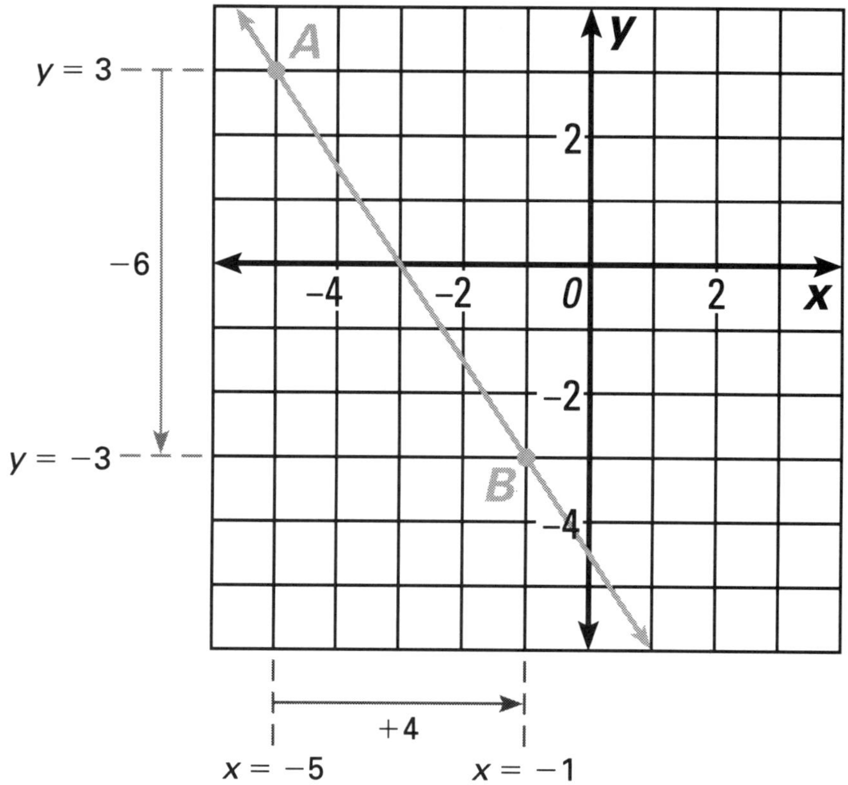

$$\text{Slope of } \overleftrightarrow{AB} = \frac{\text{rise}}{\text{run}} = \frac{\text{down 6}}{\text{right 4}} = \frac{-6}{+4} = -\frac{3}{2}$$

or

$$m = \frac{y_2 - y_1}{x_2 - x_1} = \frac{-3 - 3}{-1 - (-5)} = \frac{-6}{4} = -\frac{3}{2}$$

Graph of $y = 3x - 1$

Use $y = mx + b$.

The y–intercept, b, is −1. Plot point $(0, b)$.

The slope, m, is 3. Use it to plot two more points.

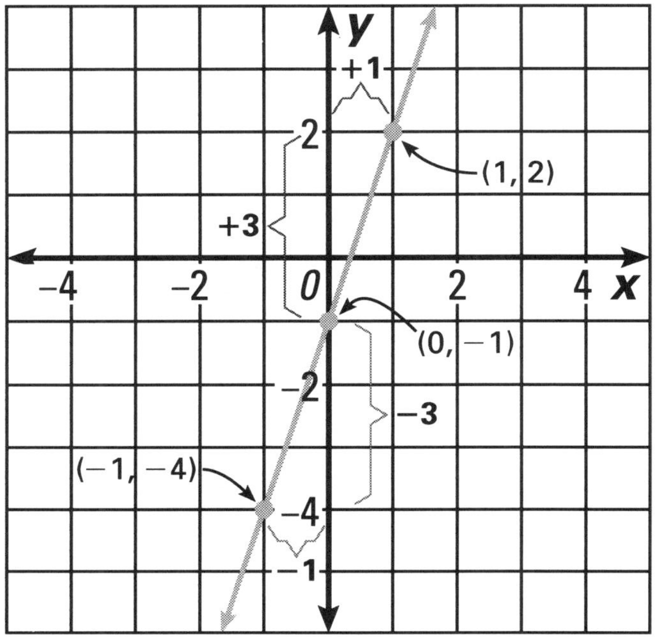

Point (1, 2) is 3 units up and 1 unit to the right of point (0, −1).

Point $(0, b)$ is point (0, −1).

Point (−1, −4) is 3 units down and 1 unit to the left of point (0, −1).

Parallel and Perpendicular Lines

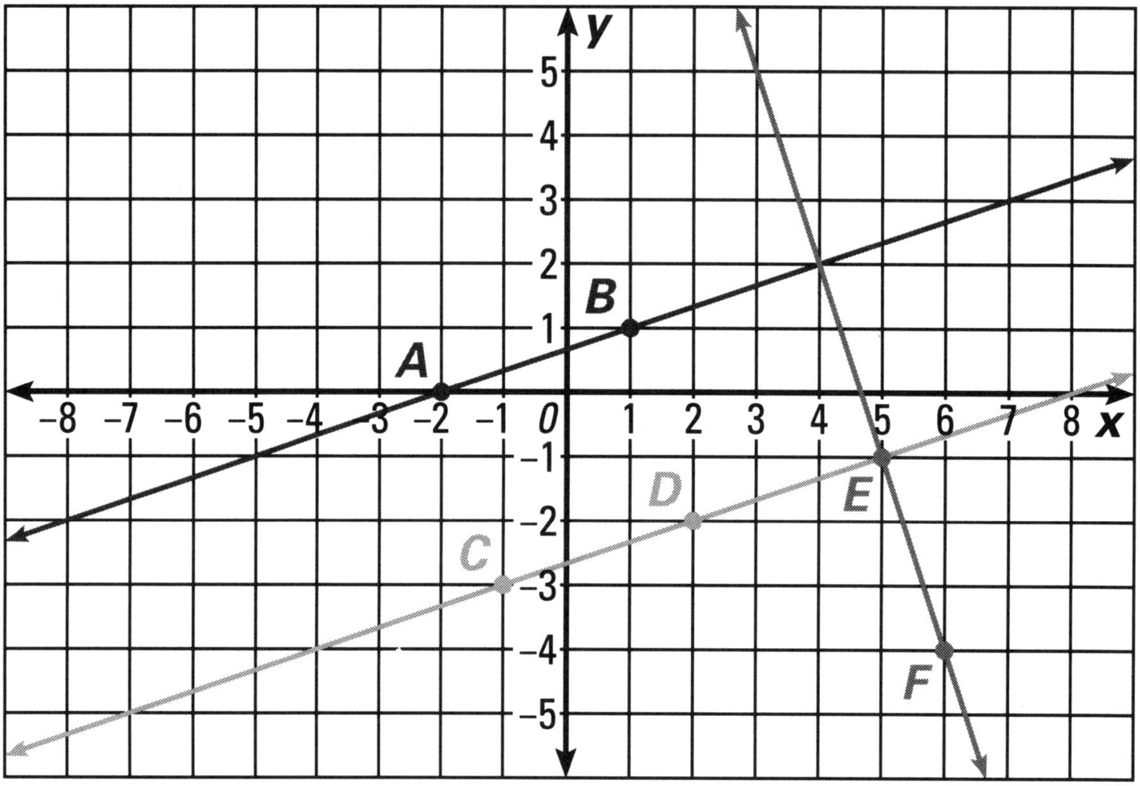

$m_{AB} = \frac{1}{3}$ $m_{CD} = \frac{1}{3}$ $m_{EF} = -3$

$m_{AB} = m_{CD} \rightarrow \overline{AB} \parallel \overline{CD}$

$(m_{AB})(m_{EF}) = -1 \rightarrow \overline{AB} \perp \overline{EF}$

Systems of Equations

Independent system
solution (1, 2)

Inconsistent system
no solution

Dependent system
infinitely many solutions

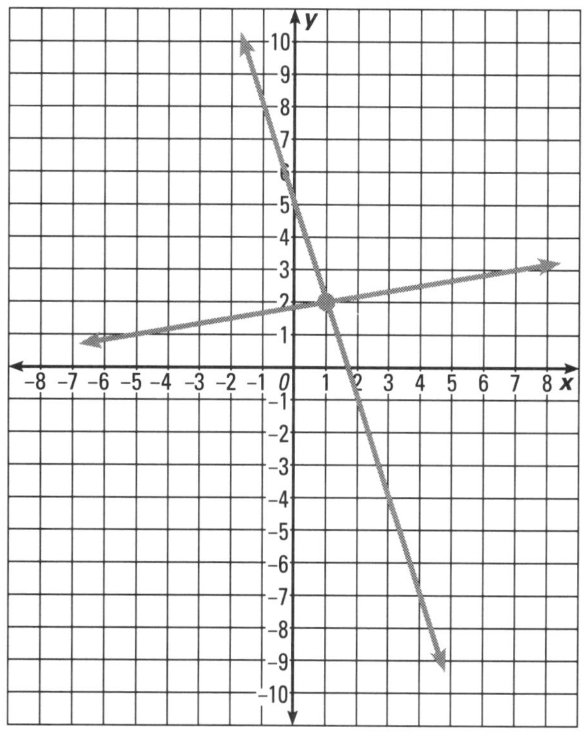

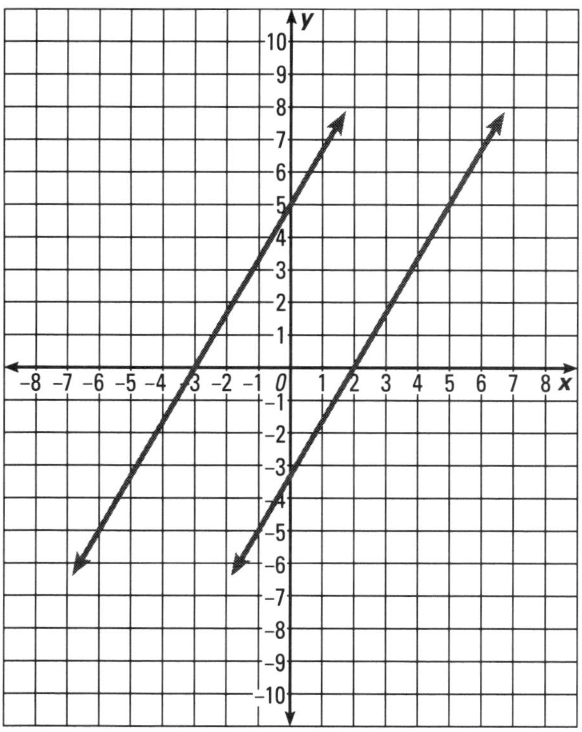

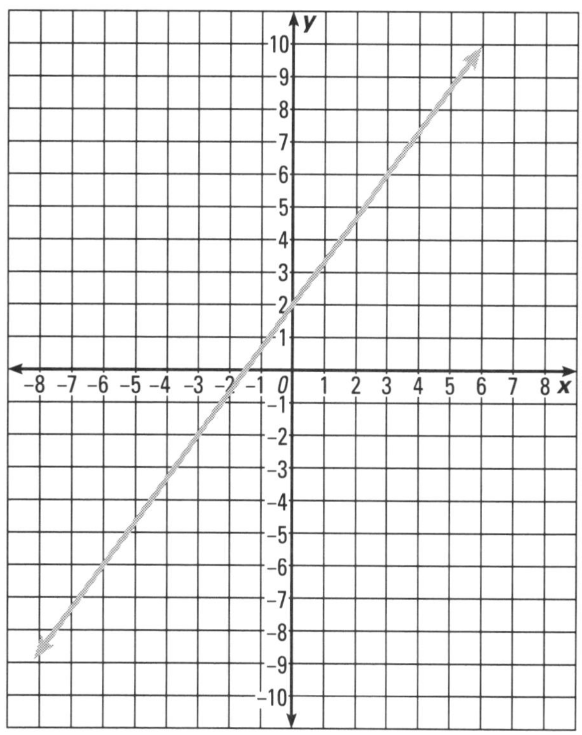

Using Determinants

System of equations → $\begin{cases} ax + by = e \\ cx + dy = f \end{cases}$ → $\det A = \begin{vmatrix} a & b \\ c & d \end{vmatrix}$ ← **Determinant of coefficients**

To find x, divide A_x, the determinant with the x-column replaced with the constant column, by determinant A.

Replace *x*-column with the constant column.

$$x = \frac{A_x}{A} = \frac{\begin{vmatrix} e & b \\ f & d \end{vmatrix}}{\begin{vmatrix} a & b \\ c & d \end{vmatrix}} = \frac{ed - bf}{ad - bc}$$

To find y, divide A_y, the determinant with the y-column replaced with the constant column, by determinant A.

Replace *y*-column with the constant column.

$$x = \frac{A_y}{A} = \frac{\begin{vmatrix} a & e \\ c & f \end{vmatrix}}{\begin{vmatrix} a & b \\ c & d \end{vmatrix}} = \frac{af - ec}{ad - bc}$$

Proportions

extreme → a, mean → c

$$\frac{a}{b} = \frac{c}{d}$$

mean → b, extreme → d

extremes → a, d

$$a:b = c:d$$

means → b, c

Product of the extremes equals the product of the means.

$$ad = bc$$

Similar Triangles

Two triangles are similar if their angles are congruent and their corresponding sides are proportional.

AA Similarity Postulate

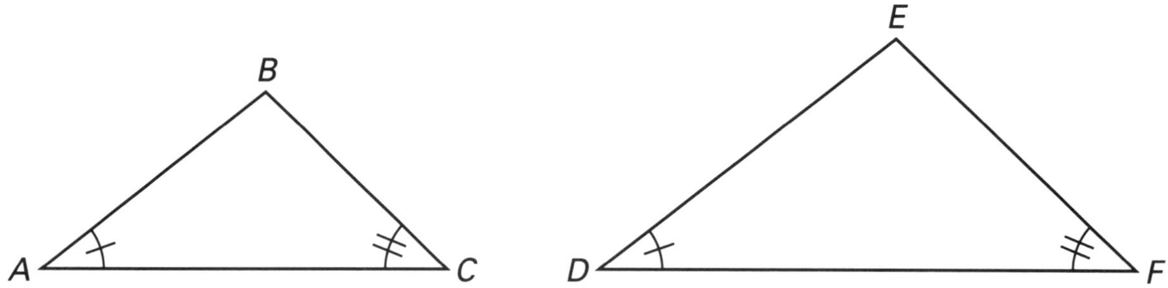

SSS Similarity Postulate

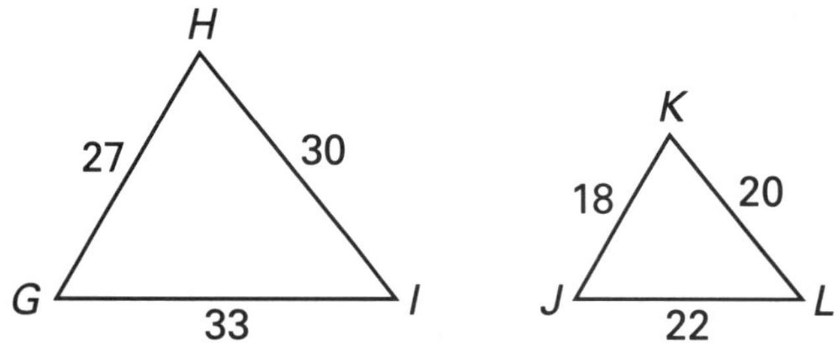

$$\frac{27}{18} = \frac{30}{20} = \frac{33}{22} = \frac{3}{2}$$

SAS Similarity Postulate

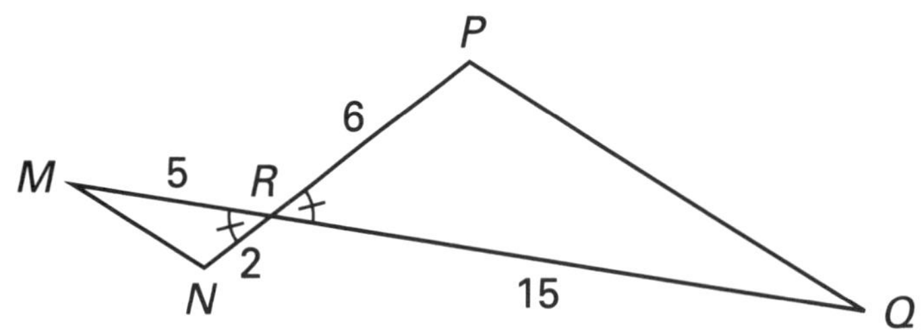

$$\frac{2}{6} = \frac{5}{15} = \frac{1}{3}$$

Translations and Reflections

MNOP ⟶ preimage

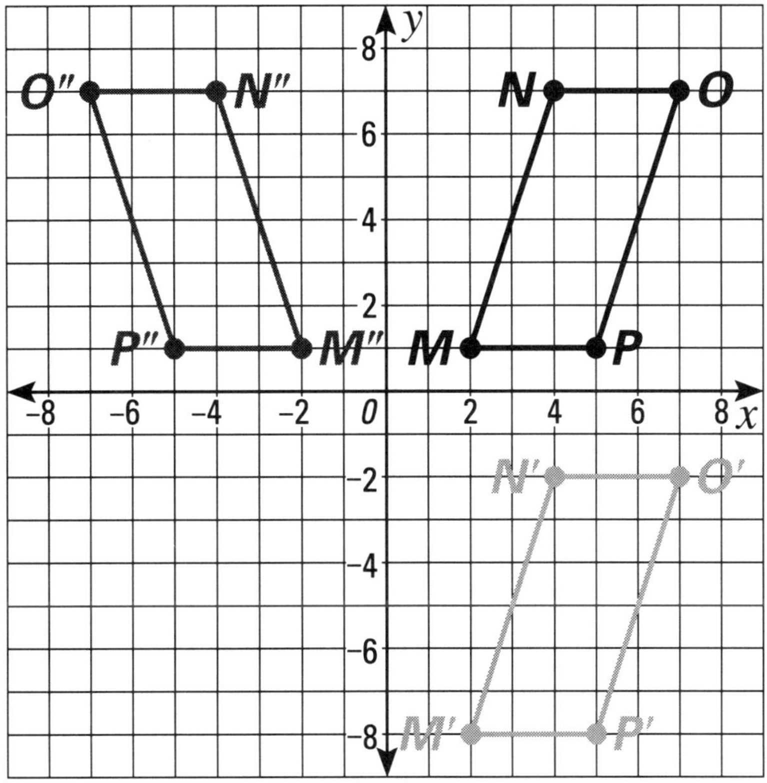

M′N′O′P′ ⟶ Translation down 9 units

M″N″O″P″ ⟶ **Reflection over the *y*-axis**

Rotations

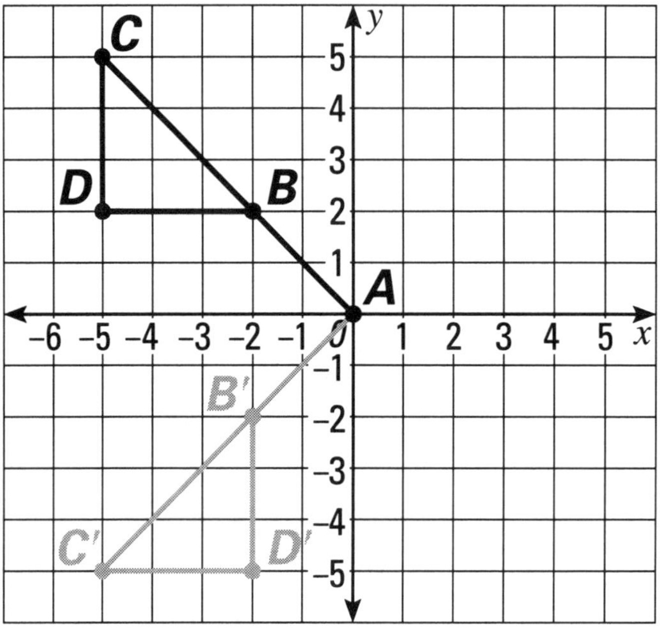
y
x
C
D
B
A
B′
C′
D′
5
4
3
2
1
−1
−2
−3
−4
−5
−6 −5 −4 −3 −2 −1 0 1 2 3 4 5

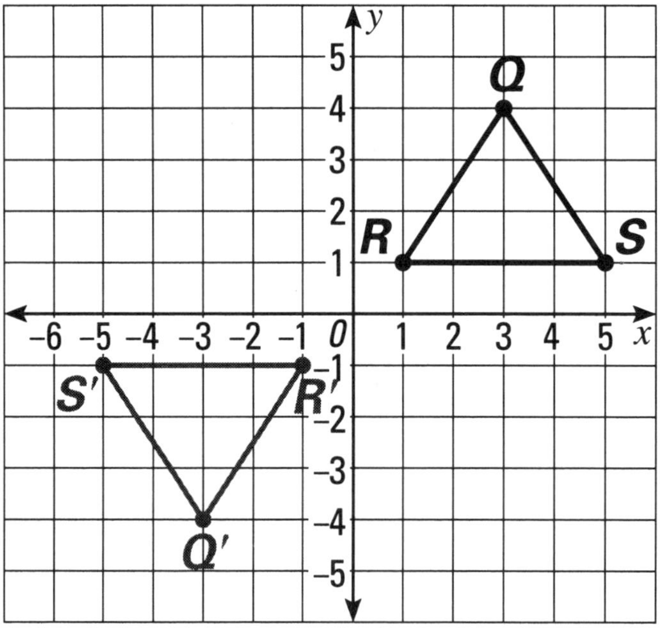
y
x
Q
R
S
S′
R′
Q′
5
4
3
2
1
−1
−2
−3
−4
−5
−6 −5 −4 −3 −2 −1 0 1 2 3 4 5

Matrix Multiplication

$$\begin{bmatrix} -1 & 5 \\ 2 & -3 \end{bmatrix} \begin{bmatrix} 3 & 0 & -3 \\ -1 & 2 & -2 \end{bmatrix}$$

The dimensions are 2 × 2 and 2 × 3, so the product will have dimensions of 2 × 3.

$$\begin{bmatrix} -1 & 5 \\ 2 & -3 \end{bmatrix} \begin{bmatrix} 3 & 0 & -3 \\ -1 & 2 & -2 \end{bmatrix} = \begin{bmatrix} -8 & \_\_ & \_\_ \\ \_\_ & \_\_ & \_\_ \end{bmatrix}$$

$(-1)(3) + (5)(-1)$

$$\begin{bmatrix} -1 & 5 \\ 2 & -3 \end{bmatrix} \begin{bmatrix} 3 & 0 & -3 \\ -1 & 2 & -2 \end{bmatrix} = \begin{bmatrix} -8 & 10 & \_\_ \\ \_\_ & \_\_ & \_\_ \end{bmatrix}$$

$(-1)(0) + (5)(2)$

$$\begin{bmatrix} -1 & 5 \\ 2 & -3 \end{bmatrix} \begin{bmatrix} 3 & 0 & -3 \\ -1 & 2 & -2 \end{bmatrix} = \begin{bmatrix} -8 & 10 & 7 \\ \_\_ & \_\_ & \_\_ \end{bmatrix}$$

$(-1)(-3) + (5)(-2)$

$$\begin{bmatrix} -1 & 5 \\ 2 & -3 \end{bmatrix} \begin{bmatrix} 3 & 0 & -3 \\ -1 & 2 & -2 \end{bmatrix} = \begin{bmatrix} -8 & 10 & 7 \\ 10 & -6 & 0 \end{bmatrix}$$

Transformations and Matrices

You can multiply matrices to find the reflection image.

Reflection	Matrix
over the x-axis	$\begin{bmatrix} 1 & 0 \\ 0 & -1 \end{bmatrix}$
over the y-axis	$\begin{bmatrix} -1 & 0 \\ 0 & 1 \end{bmatrix}$
over the line $y = x$	$\begin{bmatrix} 0 & 1 \\ 1 & 0 \end{bmatrix}$
over the line $y = -x$	$\begin{bmatrix} 0 & -1 \\ -1 & 0 \end{bmatrix}$

Spinner Patterns

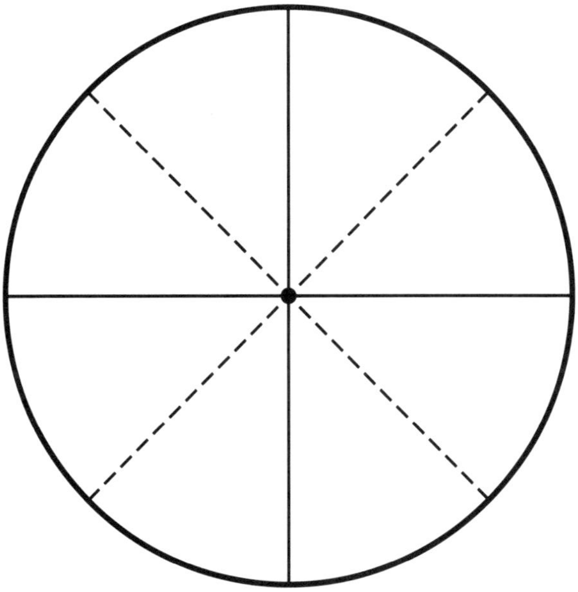

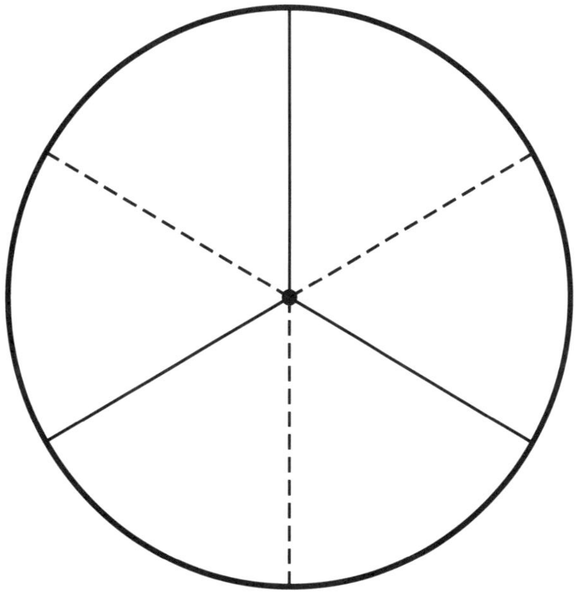

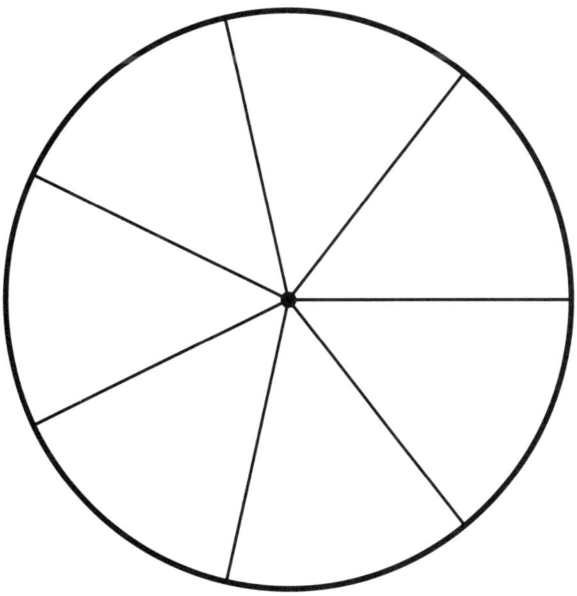

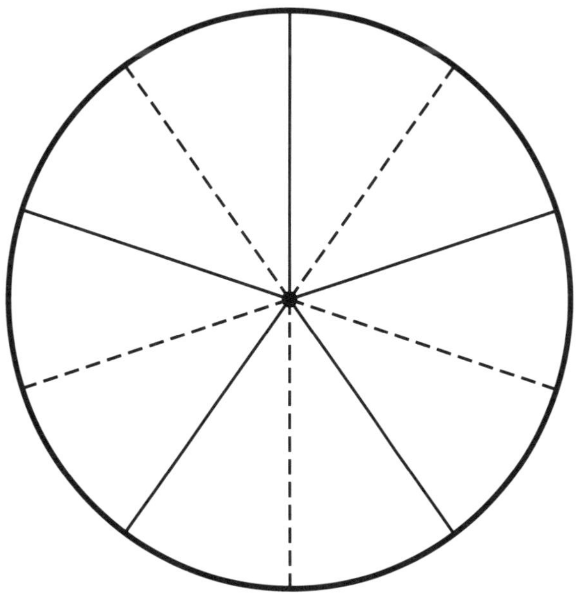

Permutations and Combinations

$$_nP_r = \frac{n!}{(n-r)!}$$

$$_nC_r = \frac{n!}{(n-r)!\,r!}$$

Scatter Plots

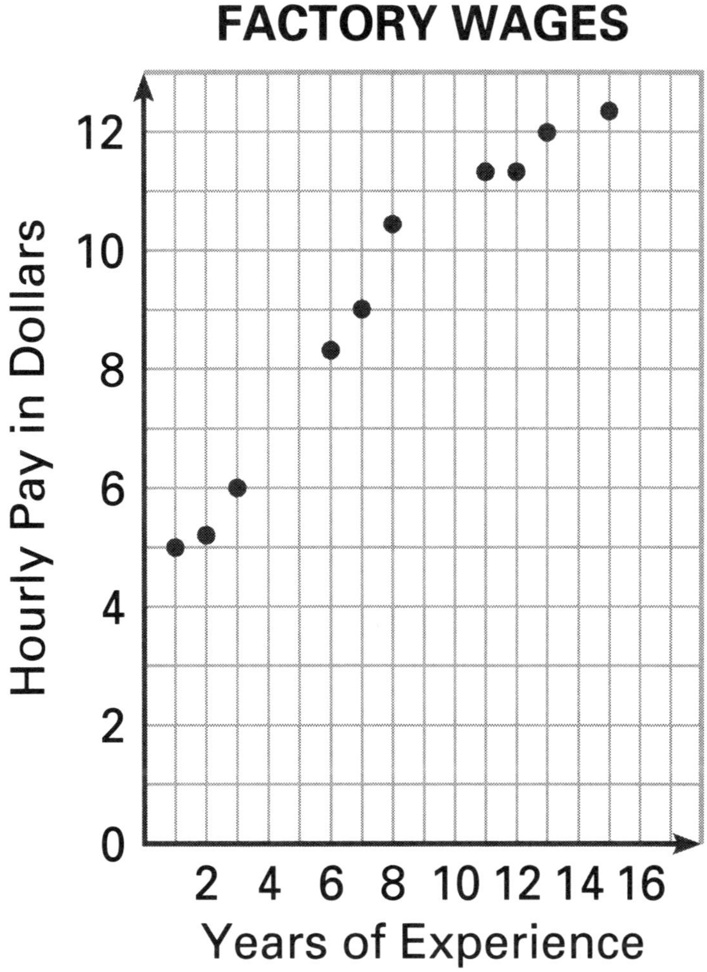

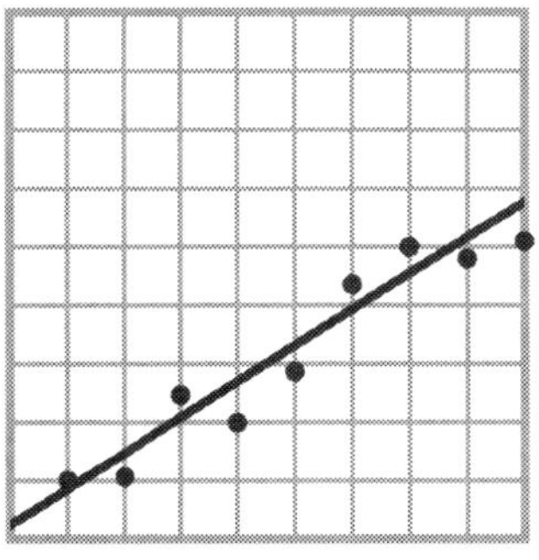

positive correlation

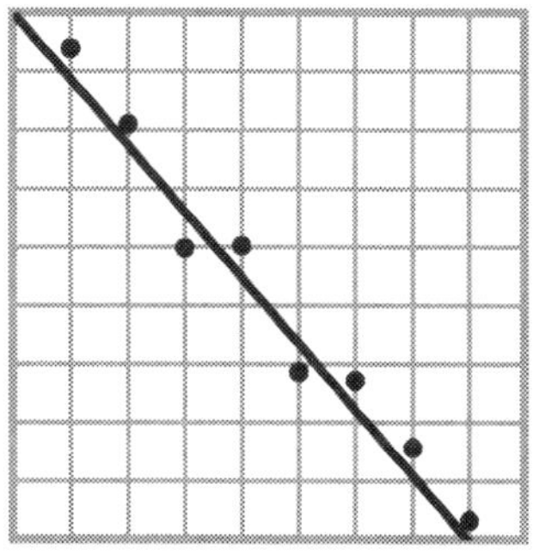

negative correlation

Box-and-Whisker Plots

least value ↓ 12, 14, 20, 21 ↑ lower quartile, 25, 29, median ↓ 30, 30, 31, 32 ↑ upper quartile, 32, 39, greatest value ↓ 46

12 14 20 21 25 29 30 30 31 32 32 39 46

DIMAGGIO'S HOME RUNS

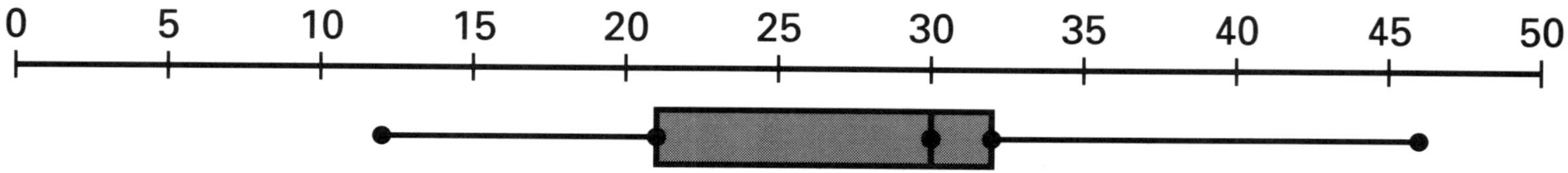

Misleading Statistics

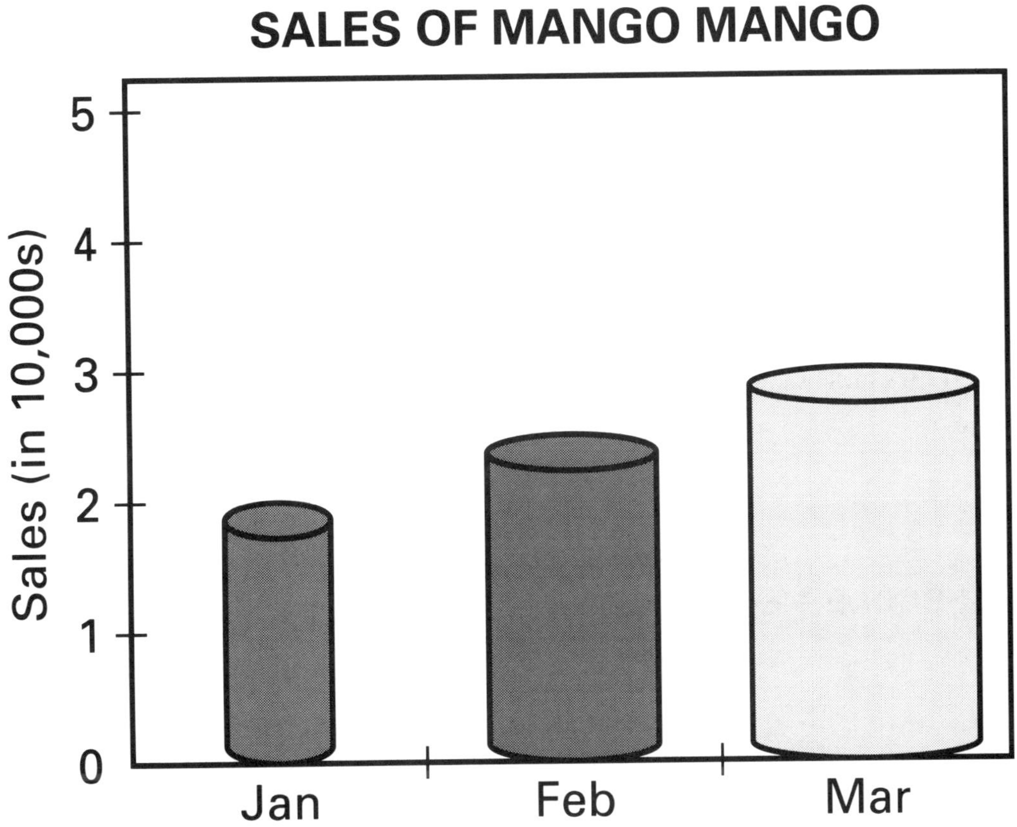

What's wrong with this circle graph?

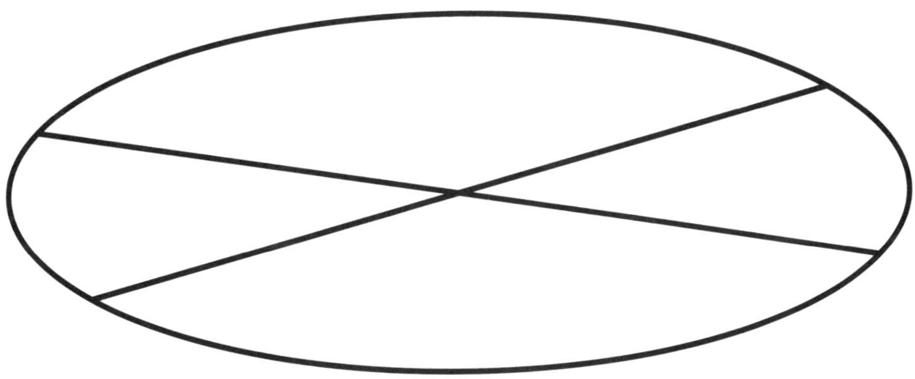

Standard Deviation

$$\text{Variance} = \frac{(x_1 - m)^2 + (x_2 - m)^2 + \ldots + (x_n - m)^2}{n}$$

$$\text{Standard Deviation} = \sqrt{\text{Variance}}$$

$$z\text{-score} = \frac{\text{score} - \text{mean}}{\text{standard deviation}}$$

Square Roots

$$\sqrt{a \cdot b} = \sqrt{a} \cdot \sqrt{b}$$

$$\sqrt{300} = \sqrt{3} \cdot \sqrt{100}$$
$$= 10\sqrt{3}$$

$$\sqrt{\frac{a}{b}} = \frac{\sqrt{a}}{\sqrt{b}}, b \neq 0$$

$$\sqrt{\frac{25}{4}} = \frac{\sqrt{25}}{\sqrt{4}}$$
$$= \frac{5}{2}$$

Rationalizing the denominator.

$$\sqrt{\frac{21}{2}} = \frac{\sqrt{21}}{\sqrt{2}}$$

$$= \frac{(\sqrt{21})(\sqrt{2})}{(\sqrt{2})(\sqrt{2})}$$

$$= \frac{\sqrt{42}}{2}$$

Pythagorean Theorem

Pythagorean Theorem

In a right triangle, the sum of the squares of the measures of the two legs is equal to the square of the measure of the hypotenuse.

$$a^2 + b^2 = c^2$$

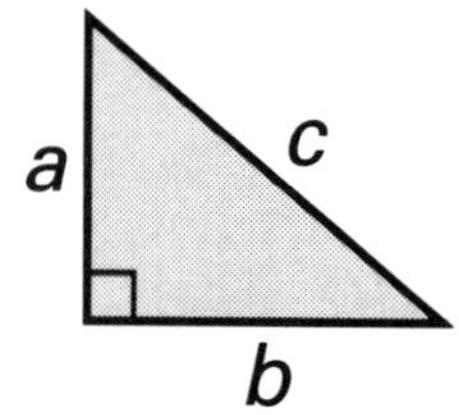

Converse of the Pythagorean Theorem

If the sum of the squares of the measures of two sides of a triangle is equal to the square of the measure of the third side, then the triangle is a right triangle.

Special Right Triangles

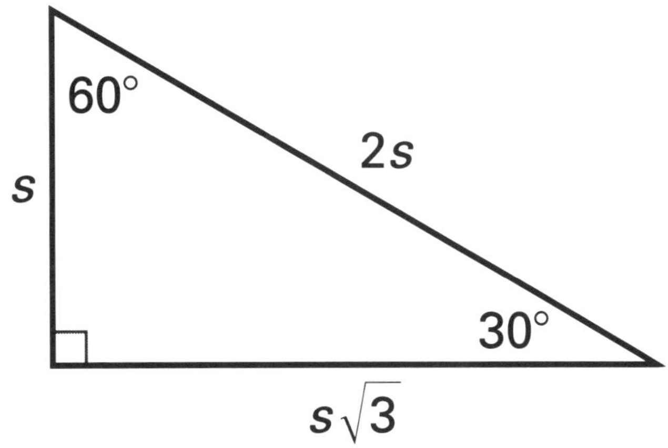

30–60–90 Triangle Theorem

In a 30–60–90 triangle, the measure of the hypotenuse is two times that of the leg opposite the 30° angle. The measure of the other leg is $\sqrt{3}$ times that of the leg opposite the 30° angle.

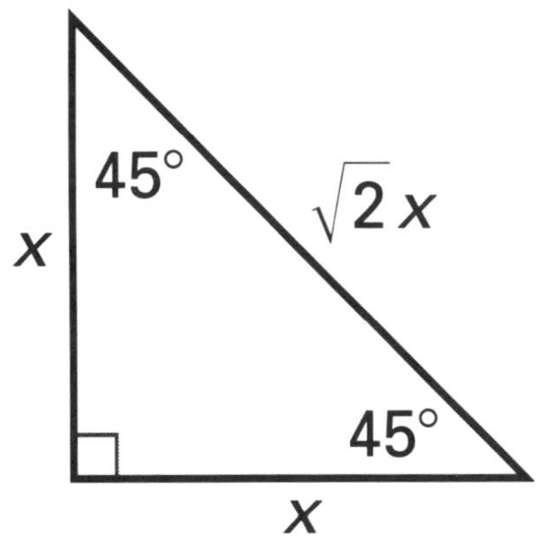

45–45–90 Triangle Theorem

In a 45–45–90 triangle, the measure of the hypotenuse is $\sqrt{2}$ times the measure of a leg of the triangle.

Circles

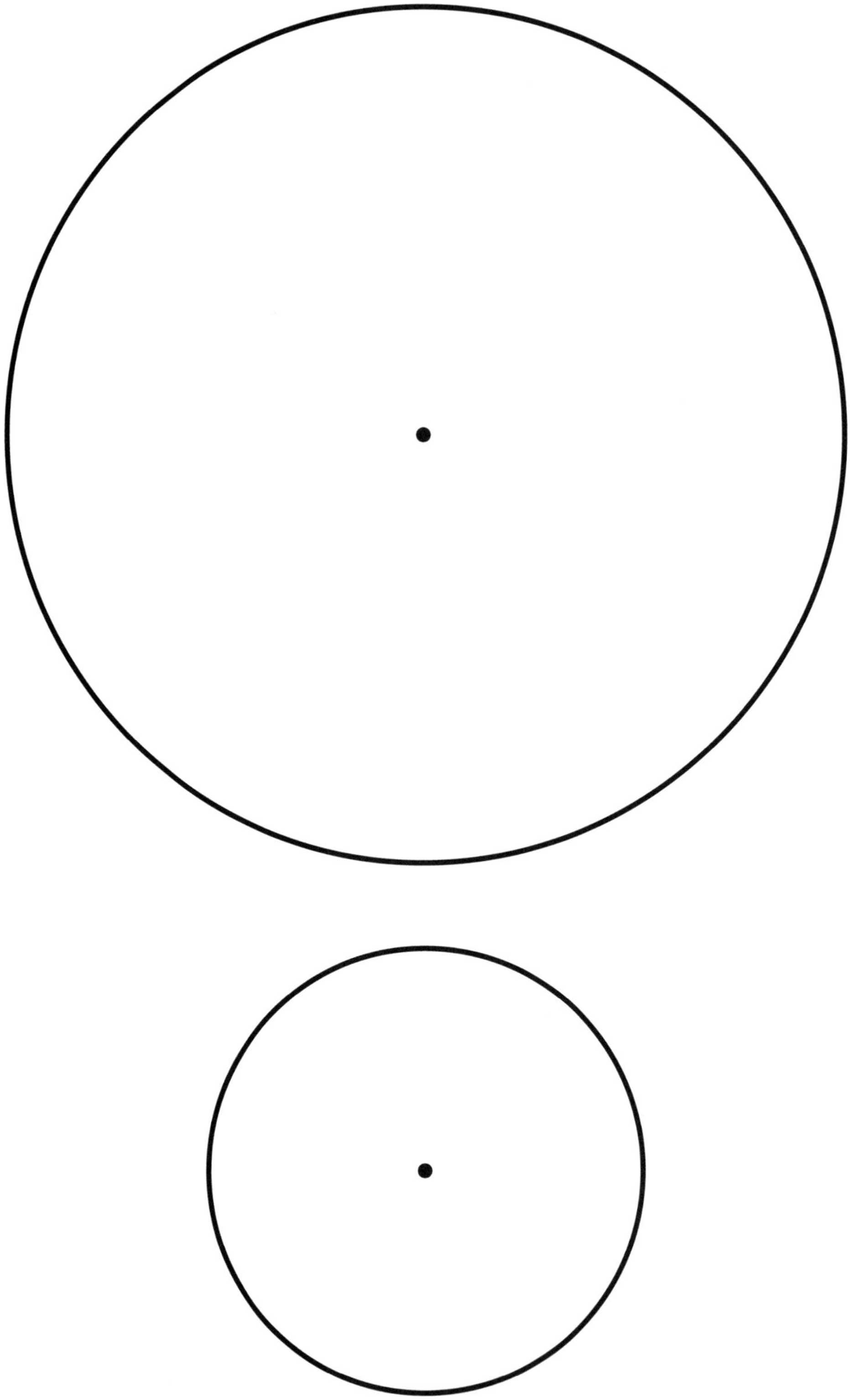

Circles, Angles, and Arcs

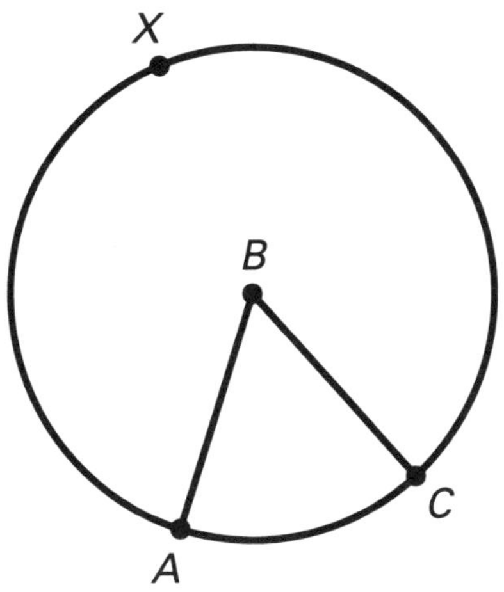

Central angle *ABC*
Minor arc *AC*: $\overset{\frown}{AC}$
Major arc *AXC*: $\overset{\frown}{AXC}$

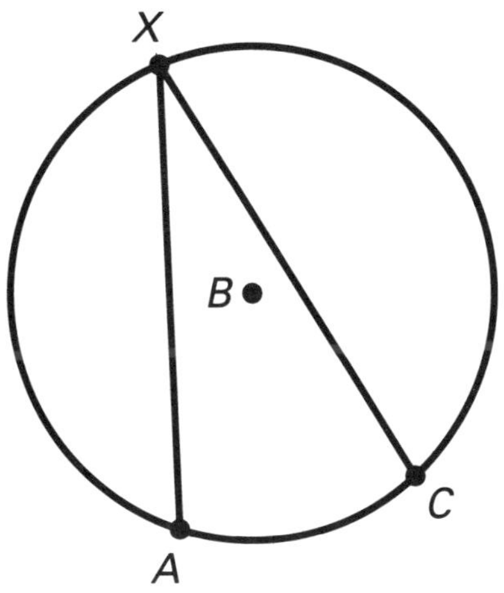

Inscribed angle *AXC*

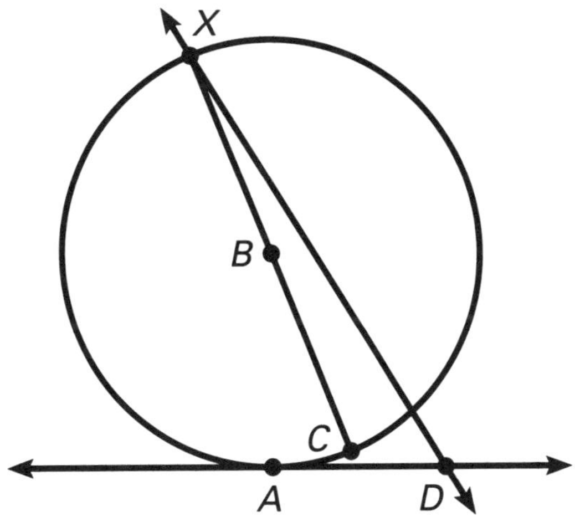

$\overline{XC}$ is a chord.
$\overline{XC}$ is a diameter.
$\overleftrightarrow{XD}$ is a secant.
$\overleftrightarrow{AD}$ is a tangent.

Circles and Segments

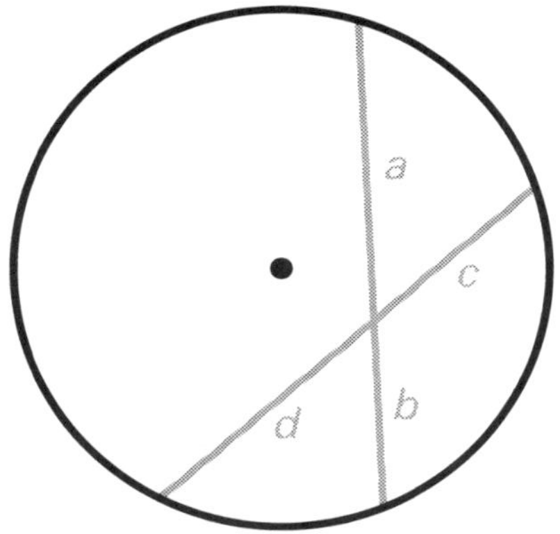

$ab = cd$

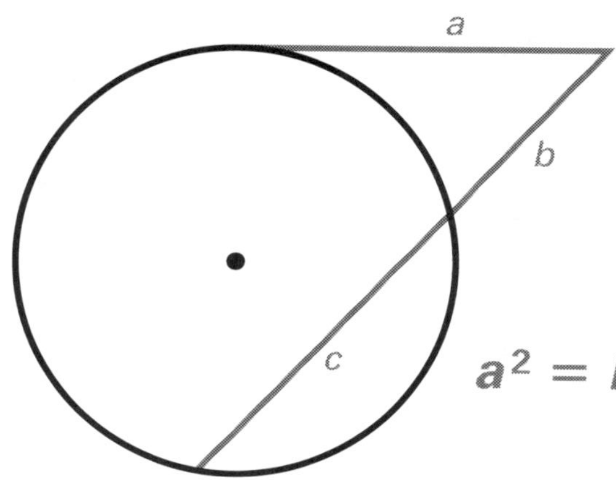

$a^2 = b(b + c)$

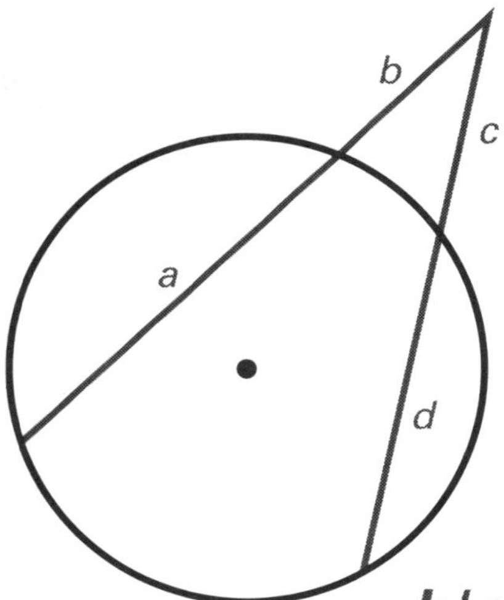

$b(a + b) = c(c + d)$

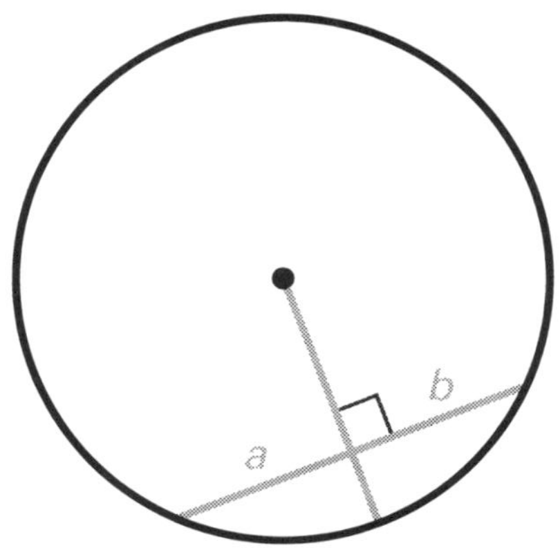

$a = b$

FOIL Process

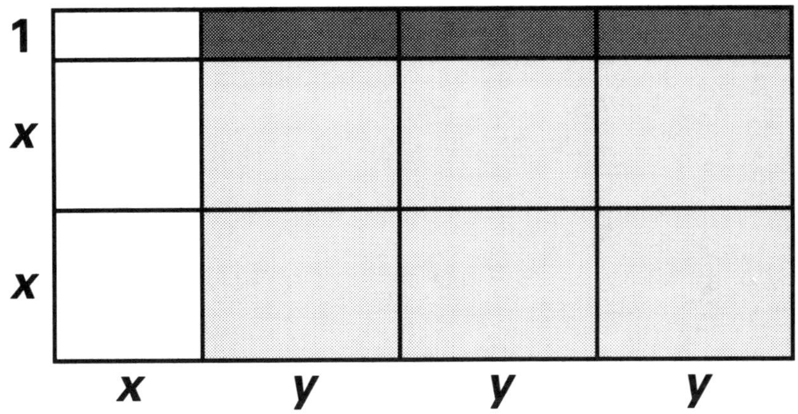

$(2x + 1)\ (x + 3y)$

F	(first)	$(2x + 1)\ (x + 3y)$	$2x^2$
O	(outer)	$(2x + 1)\ (x + 3y)$	$6xy$
I	(inner)	$(2x + 1)\ (x + 3y)$	x
L	(last)	$(2x + 1)\ (x + 3y)$	$3y$

Add.

$(2x + 1)\ (x + 3y) = 2x^2 + x + 6xy + 3y$

Special Factoring Patterns

	Pattern of trinomial	How it relates to binomial
First term	A perfect square	The square of the binomial's first term
Last term	A perfect square	The square of the binomial's last term
Middle term	The square roots of the two perfect squares multiplied together, and then doubled	The product of the binomial's terms, multiplied by two

Graphing Parabolas

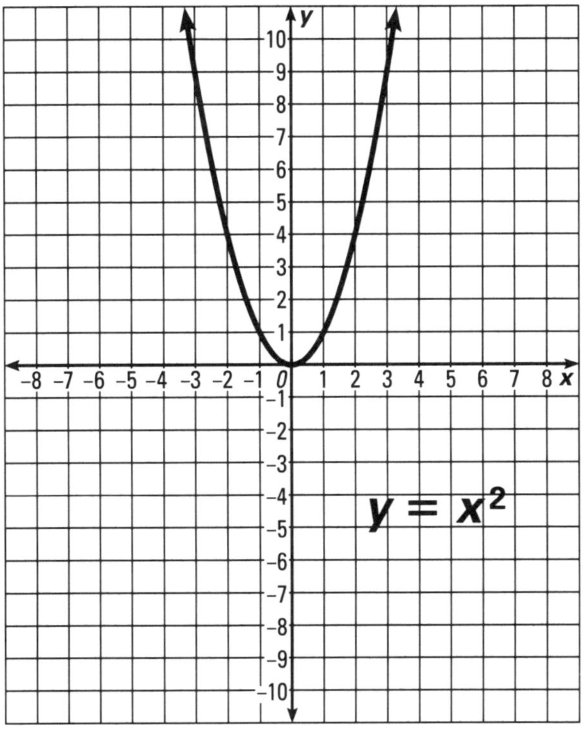

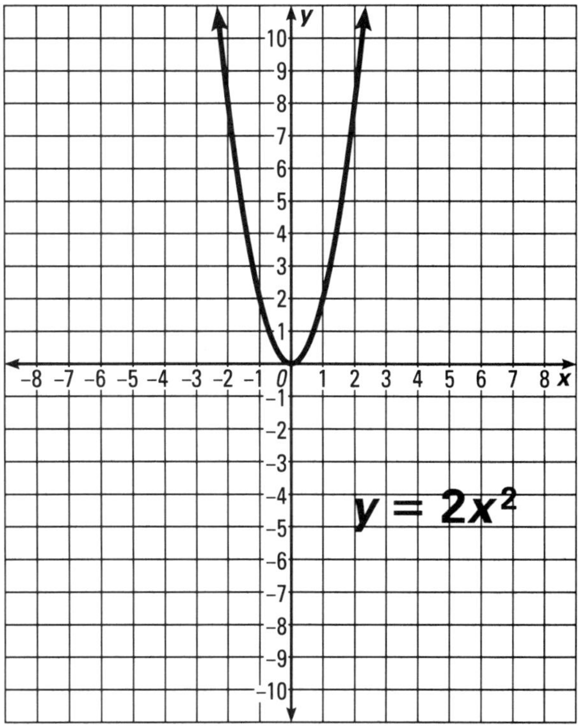

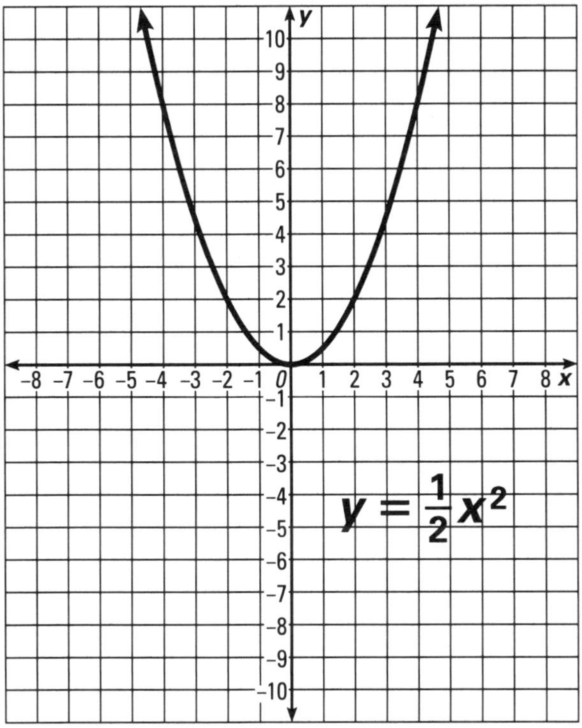

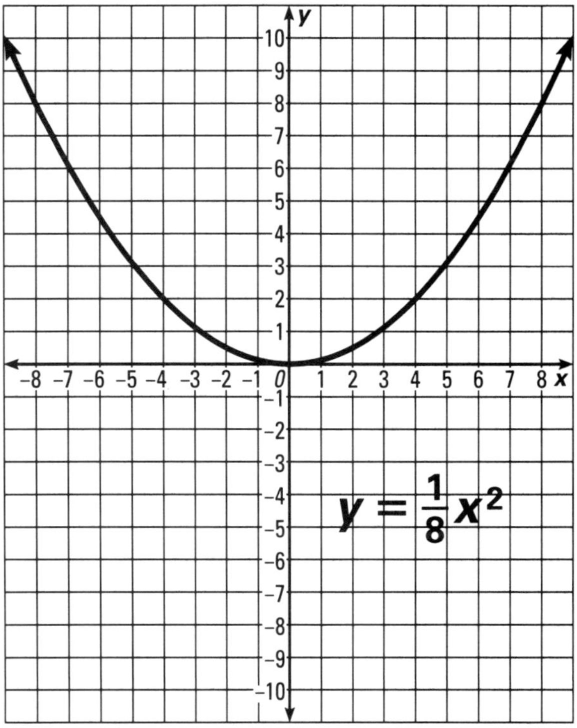

Graphing Parabolas

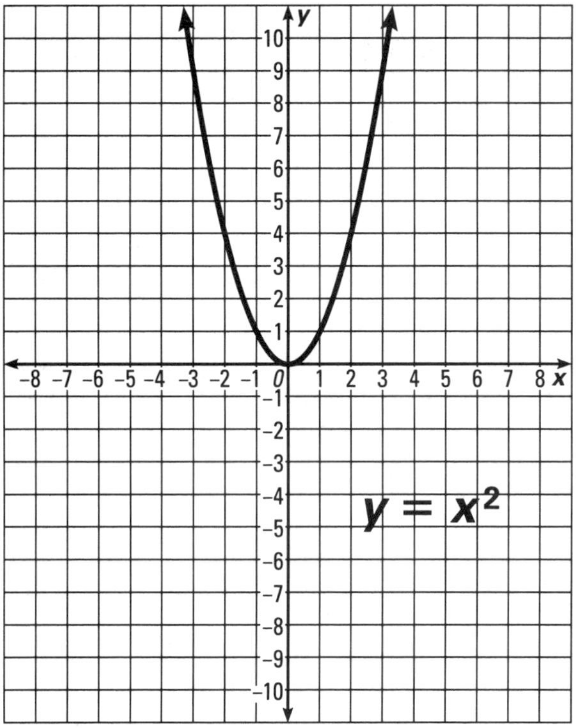

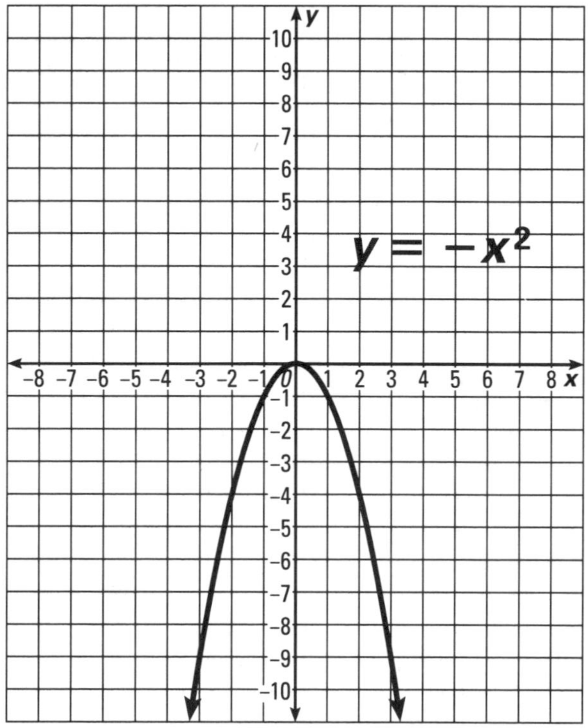

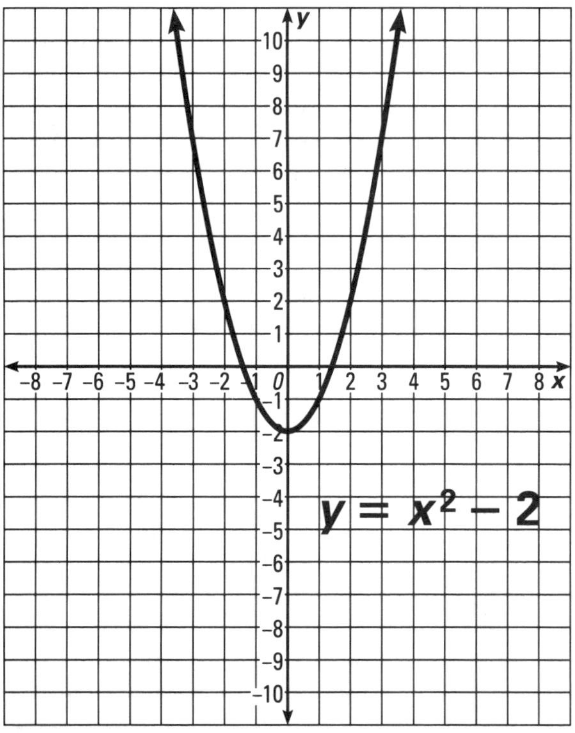

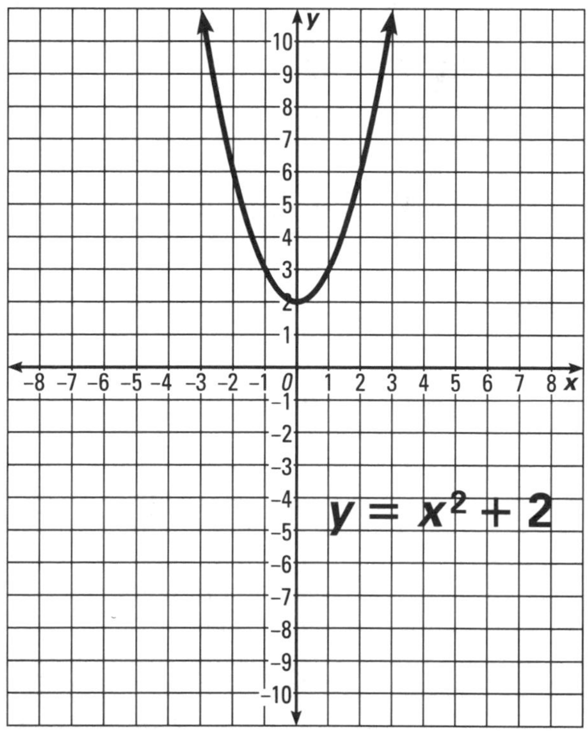

The Quadratic Formula

$$ax^2 + bx + c = 0$$

$$x^2 + \frac{b}{a}x + \frac{c}{a} = 0$$ ← Multiply each term by $\frac{1}{a}$.

$$x^2 + \frac{b}{a}x = -\frac{c}{a}$$ ← Add $-\frac{c}{a}$ to each side of the equation.

$$x^2 + \frac{b}{a}x + \frac{b^2}{4a^2} = \frac{b^2}{4a^2} - \frac{c}{a}$$ ← Half of $\frac{b}{a} = \frac{b}{2a}$. Add $\left(\frac{b}{2a}\right)^2$ or $\frac{b^2}{2a^2}$ to each side of the equation.

$$\left(x + \frac{b}{2a}\right)^2 = \frac{b^2 - 4ac}{4a^2}$$ ← Factor the left side of the equation. Combine terms on the right side of the equation.

$$x + \frac{b}{2a} = \pm\sqrt{\frac{b^2 - 4ac}{4a^2}}$$ ← Find the square roots.

$$x = -\frac{b}{2a} \pm \frac{\sqrt{b^2 - 4ac}}{2a}$$ ← Subtract $\frac{b}{2a}$ from each side of the equation.

$$x = \frac{-b \pm \sqrt{b^2 - 4ac}}{2a}$$ ← Simplify

Using the Pythagorean Theorem

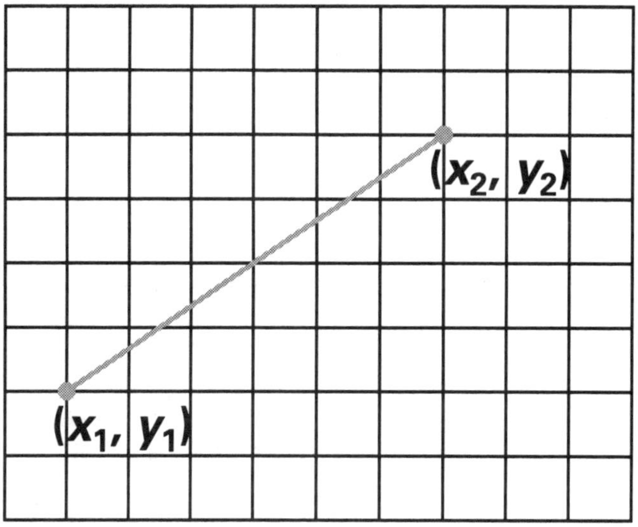

Distance Formula

$$d = \sqrt{(x_2 - x_1)^2 + (y_2 - y_1)^2}$$

Midpoint Formula

$$\text{Midpoint} = \left(\frac{x_1 + x_2}{2}, \frac{y_1 + y_2}{2}\right)$$

Standard Equation of a Circle

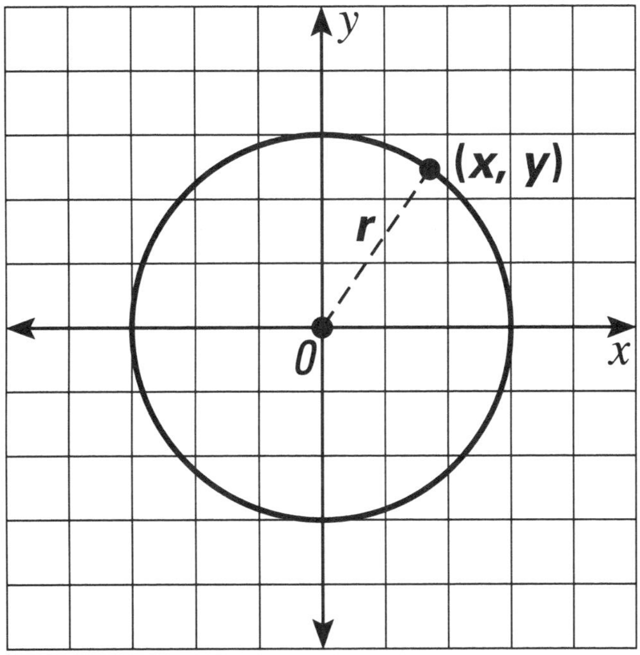

Center at the origin: $x^2 + y^2 = r^2$

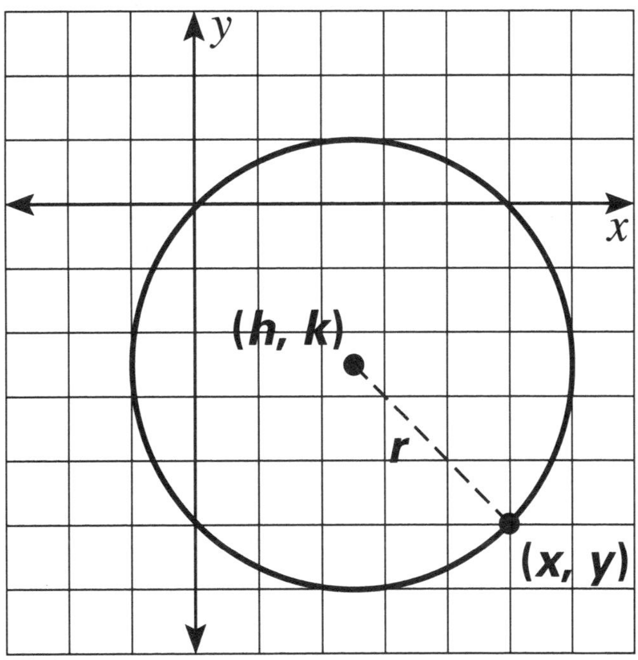

Center at (h, k): $(x - h)^2 + (y - k)^2 = r^2$

Ellipses and Hyperbolas

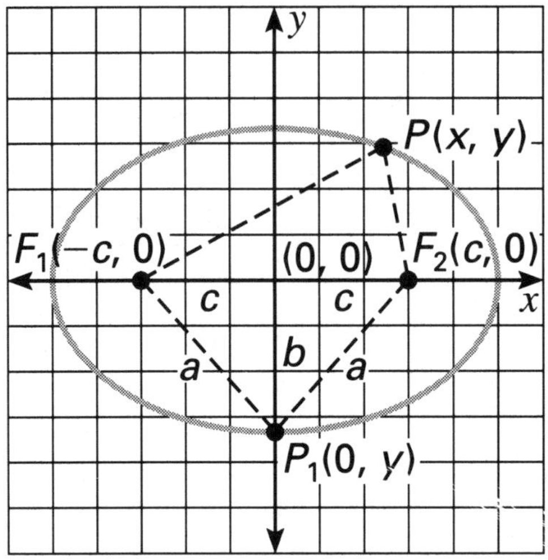

$$\frac{x^2}{a^2} + \frac{y^2}{b^2} = 1$$

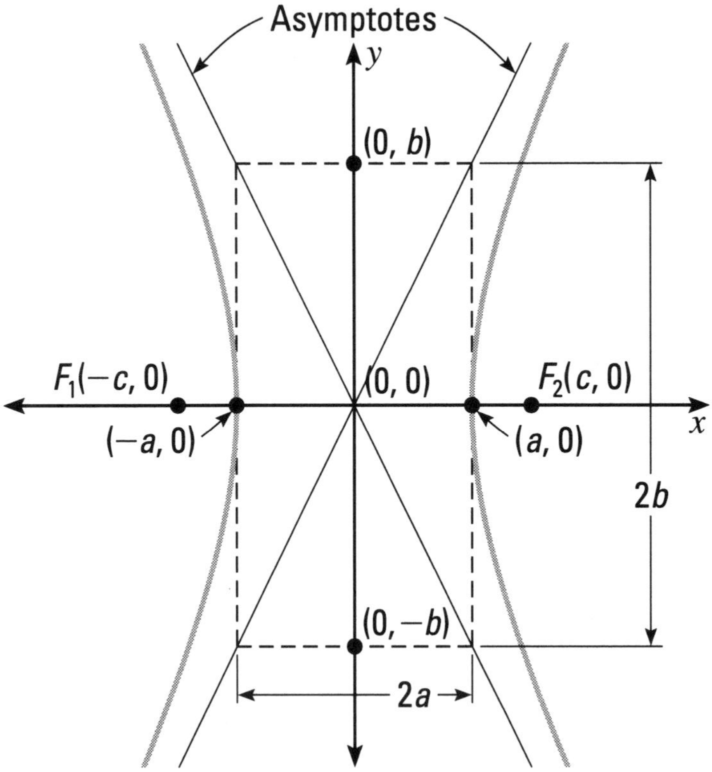

$$\frac{x^2}{a^2} - \frac{y^2}{b^2} = 1$$

Direct Variation

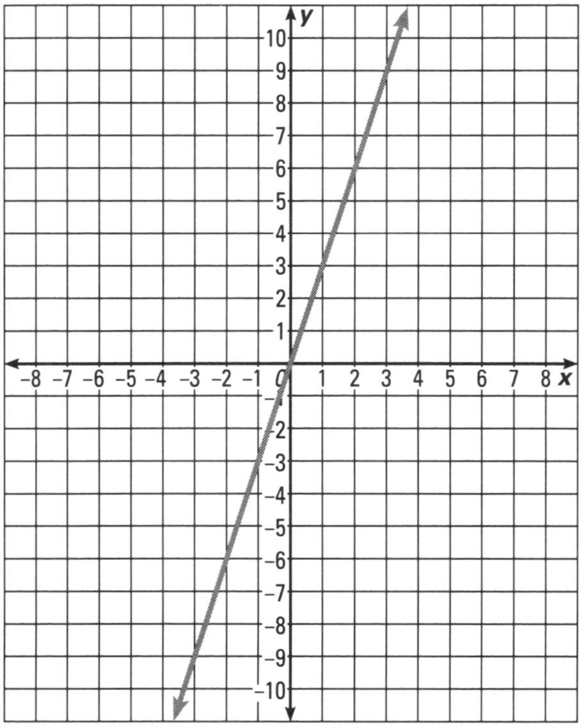

Direct Variation

$y = 3x$

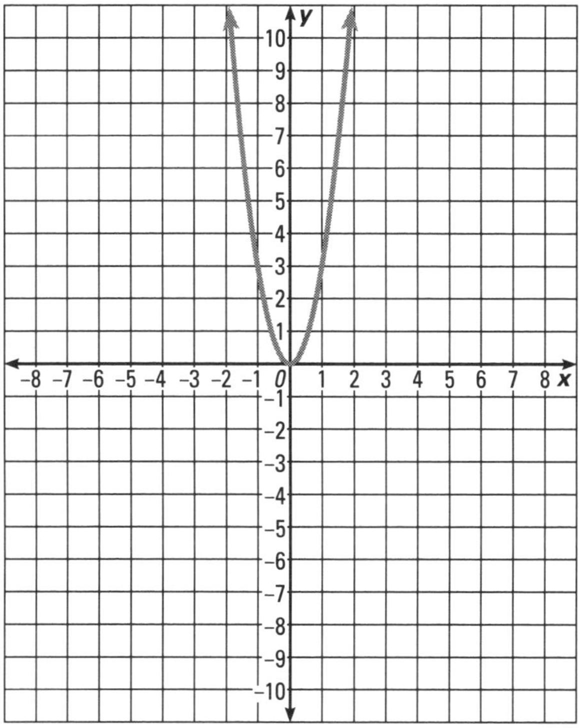

Direct Square Variation

$y = 3x^2$

Inverse Variation

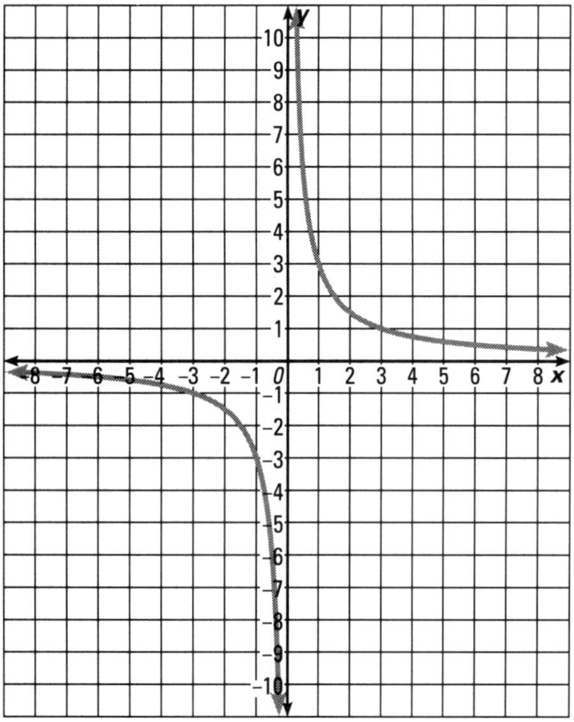

Inverse Variation

$$y = \frac{3}{x}$$

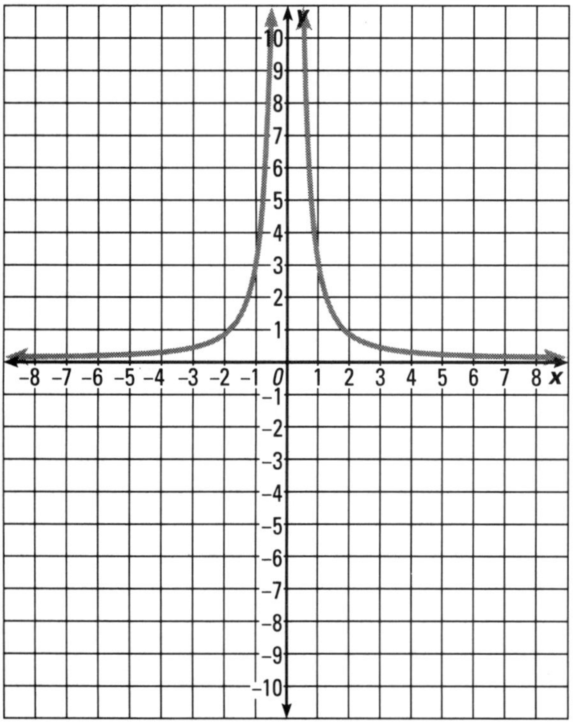

Inverse Square Variation

$$y = \frac{3}{x^2}$$

Right Triangle Relationships

In relation to $\angle R$:
$\overline{RT}$ is the side adjacent to $\angle R$.
$\overline{ST}$ is the side opposite $\angle R$.

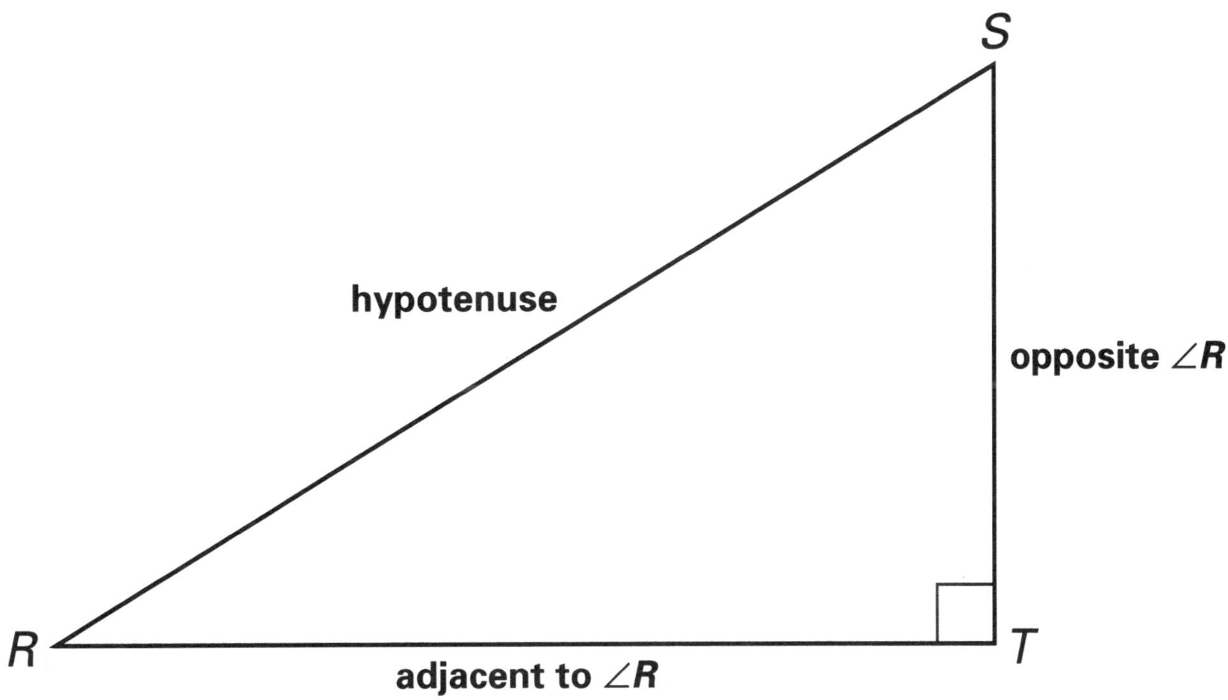

Trigonometric Ratios for Right Triangle *ABC*

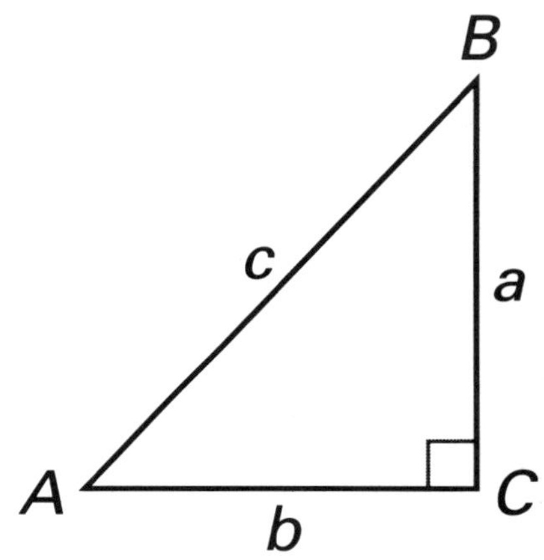

Name of ratio	Abbreviation	Ratio
sine of $\angle A$	sin A	$\dfrac{\text{length of side opposite } \angle A}{\text{hypotenuse}}$ or $\dfrac{a}{c}$
cosine of $\angle A$	cos A	$\dfrac{\text{length of side adjacent to } \angle A}{\text{hypotenuse}}$ or $\dfrac{b}{c}$
tangent of $\angle A$	tan A	$\dfrac{\text{length of side opposite } \angle A}{\text{length of side adjacent to } \angle A}$ or $\dfrac{a}{b}$

Solving Right Triangles

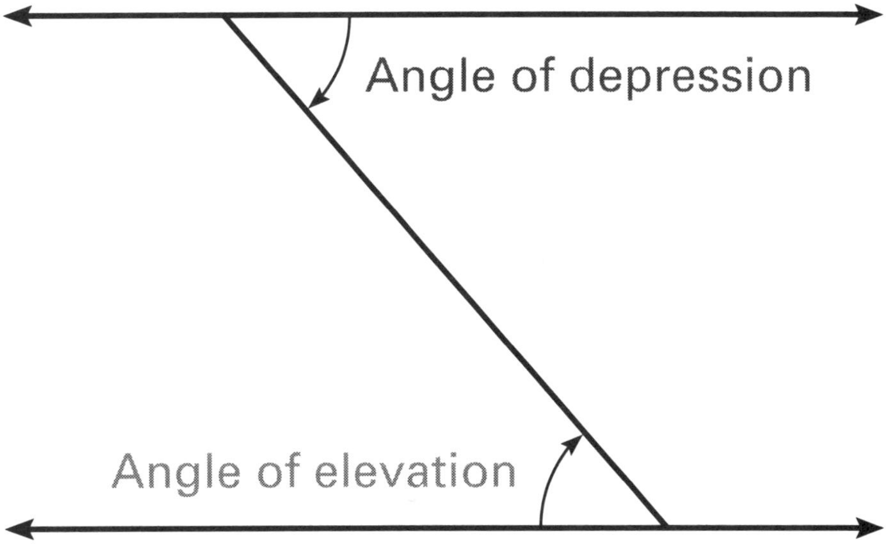

Reference Angle

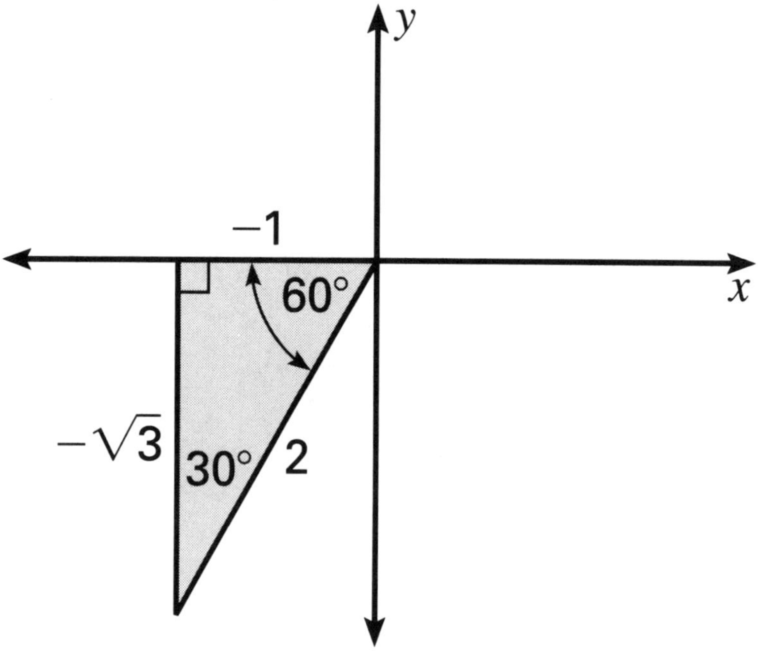

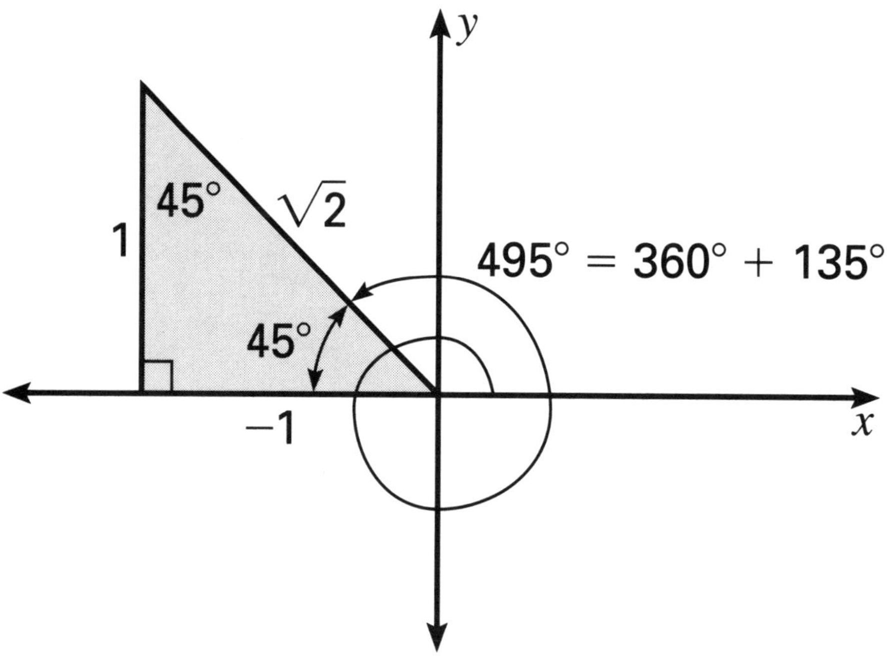